自主创新 方法先行

本书的出版受科技部创新方法项目（2008IM020200）支持

当代科学文化前沿丛书

METHODOLOGY OF
THE HISTORY OF SCIENCE

科学史方法论讲演录

〔美〕席文 ◎著
任安波 ◎译
任定成 ◎校

北京市版权局著作权合同登记号　01-2011-4881
图书在版编目(CIP)数据

科学史方法论讲演录/(美)席文(Sivin,N.)著;任安波译;任定成校.—北京:北京大学出版社,2011.11
(当代科学文化前沿丛书)
ISBN 978-7-301-19689-2

Ⅰ.①科… Ⅱ.①席…②任…③任… Ⅲ.①科学史－方法论－文集 Ⅳ.①G3-53

中国版本图书馆 CIP 数据核字(2011)第 225646 号

Methodology of the History of Science / by Nathan Sivin
Peking University Press is Granted the right to print, publish and sell the English-Chinese edition throught the world.
This license is free of charge.
The termination of the license is subject to a mutual agreement between Peking University Press and Nathan Sivin.

书　　　名:科学史方法论讲演录
著作责任者:[美]席　文　著　任安波　译　任定成　校
策 划 编 辑:周雁翎
责 任 编 辑:周志刚
标 准 书 号:ISBN 978-7-301-19689-2/N·0044
出 版 发 行:北京大学出版社
地　　　址:北京市海淀区成府路 205 号　100871
网　　　站:http://www.jycb.org　http://www.pup.cn
电 子 信 箱:zyl@pup.pku.edu.cn
电　　　话:邮购部 62752015　发行部 62750672　编辑部 62767346
　　　　　　出版部 62754962
印　刷　者:三河市博文印刷厂
经　销　者:新华书店
　　　　　　650 毫米×980 毫米　16 开本　13.75 印张　141 千字
　　　　　　2011 年 11 月第 1 版　2011 年 11 月第 1 次印刷
定　　　价:30.00 元

未经许可,不得以任何方式复制或抄袭本书之部分或全部内容。
版权所有,侵权必究
举报电话:(010)62752024　电子信箱:fd@pup.pku.edu.cn

献给已故的竺可桢、席泽宗、陈美东、李迪和博树人诸位教授

目　录

序 ··· (1)

第一讲　科学史和医学史正发生着怎样的变化 ············· (7)
　　职业科学史 ··· (10)
　　一般科学史的发展 ··· (11)
　　焦点变化 ··· (16)
　　对中国科学史的新探讨 ··································· (18)
　　使用新工具 ··· (19)
　　下一步 ··· (21)

第二讲　运用社会学和人类学方法 ························· (23)
　　人类学 ··· (26)
　　人类学与医学 ·· (28)
　　社会学 ··· (33)
　　医学社会学 ··· (34)
　　疾病社会学 ··· (35)
　　结语 ·· (38)

第三讲　运用大众文化研究方法 ···························· (39)
　　精英医学的宇宙观 ··· (43)
　　大众医学 ·· (44)
　　效力 ··· (47)

结语 ··· (52)

第四讲　比较 ··· (55)
　　　先决条件 ··· (57)
　　　比较的例子 ·· (59)
　　　自然 ··· (60)
　　　身体 ··· (61)
　　　东亚使用的天文学和医学 ·· (63)
　　　网络技术 ··· (66)
　　　结语 ··· (70)

第五讲　运用"文化簇"概念 ··· (71)
　　　伊斯兰对中国天文学的影响 ······································· (75)
　　　博学之士 ··· (79)
　　　"李约瑟问题" ·· (82)
　　　不二臣 ··· (85)
　　　结语 ··· (87)

人名译名表 ··· (89)

译后记 ··· (93)

CONTENTS

Preface .... (97)

1 How the History of Science and Medicine is Changing ...... (105)

    Professional History of Science .... (108)

    Developments in General History of Science .... (110)

    A Change of Focus .... (115)

    New Initiatives in the Study of Chinese Science .... (117)

    New Tools to Apply .... (118)

    Further Steps .... (120)

    References .... (121)

2 Using the Methods of Sociology and Anthropology ...... (125)

    Anthropology .... (128)

    Anthropology and medicine .... (131)

    Sociology .... (136)

    Sociology of Medicine .... (138)

    Sociology of Disease .... (140)

    Conclusion .... (142)

    References .... (143)

3 Using Studies of Popular Culture .... (145)

    The world view of elite medicine .... (150)

  Popular Medicine ··· (151)
  Efficacy ··· (155)
  Conclusion ··· (161)
  References ··· (163)

4 Using Comparison ··· (165)
  Prerequisites ··· (167)
  Examples of Comparison ······································· (169)
  Nature ··· (170)
  Body ·· (172)
  East Asian Uses of Astronomy and Medicine ············ (174)
  Networks ··· (178)
  Conclusion ··· (182)
  References ··· (183)

5 Using Cultural Manifolds ······································ (185)
  Islamic Influence on Chinese Astronomy ················· (189)
  Polymaths ·· (194)
  The "Needham Question" ······································ (197)
  Bu'erchen ·· (201)
  Conclusion ··· (203)
  References ··· (205)

序

很荣幸应中国科学院之邀,发表2009年竺可桢讲演。为了鼓励学生参与,这些讲演于4月13日至22日在北京大学进行。①

由于1977年才能访问中国,因此我从未见过生活在1890年至1974年的竺可桢教授。但是我早就知道他的工作,因为他的著作对于每个研习中国科学史的学生来说都是必读著作。他就范围非常广泛的历史问题撰写著作,从地质学、气象学到天文学。他1926年发表的对于沈括的研究,启发我自己撰写关于沈括的著作。他1954年发表在《人民日报》上、题为"为什么要研究我国古代科学史?"的文章,鼓舞了许多读者。② 简言之,他是我由之学习历史研究方法的许多现代中国学者之中的第一位。

这就是我的这些演讲专门讲述历史研究方法的原因。我花了50年时间阅读、思考和撰写中国科学史。③ 其间,一般的历史探究方法发生了根本变化。这些变化多数发生在史学家们使其他学科的工具适应他们自己的需要之时。半个世纪之前,多数

---

① 感谢任定成教授和北京大学科学与社会研究中心的成员以及中国科学院自然科学史研究所孙小淳研究员在这些讲演的准备、实施和出版中所给予的慷慨帮助。也感谢自然科学史研究所郑术给予的帮助和建议。

② 竺可桢,"北宋沈括对于地学之贡献与纪述",《科学》,1926,第11卷第6期,第792—807页;竺可桢,"为什么要研究我国古代科学史?",《人民日报》,1954年8月27日第3版。

③ 这里和接下来的各讲中,我说的"科学史"包括技术史与医学史。限于篇幅,本书将不讨论技术史。

所谓智识史依赖的是哲学分析方法。二十世纪60年代和70年代,人口统计数据变得有用起来,接着就是经济学、社会学和人类学。最近十年左右,环境研究方法导致了环境史研究。

20世纪70年代以前,社会科学的运用使主导的史学家们的研究重点从智识史逐步变为社会史。到80年代,我的许多研究一般历史的同仁认识到,社会史和智识史同样太狭隘,容纳不了历史复杂性。结果,同时从多种视角看待问题的兴趣日益增长。文化史考察一个历史时期的艺术、物质文化、经济和社会建制,不过研究文化史的人对其定义并没有达成一致。第五讲描述的称为"文化簇"的较新进路稍微往前延伸了一点,它首先确定所有相关维度来考察问题,然后把全部这些维度作为单个模式的部分来探究。近来其他同样重要的研究鼓舞着学者们避免假定他们所研究的团体的所有成员意见一致,激励学者们承认社会决策是不同个体之间的歧见和冲突的妥协结果。

这些新发展惊人地改变了一般的历史。它们也影响了最有进取心的科学史家们,并逐渐影响到他们的学生。部分由于这些新进路,物质科学史家们比以前更倾向于探究诸如科学欺骗、剽窃、与公共利益的冲突、公众和政府在科学资助中的作用以及科学家自我推销之类的问题。技术史家不再把过去的技术看作是应用科学的一部分;19世纪以前的工匠很少接受他们时代的科学训练,他们传承的是独一无二的技能。

在许多国家,医学史是分开的。一些史学家,主要是那些在医学院任教的史学家,仍然把他们的领域看作是进步编年史,很少关注非医生的治疗。其他史学家,主要是那些历史系和科学史系的史学家,更倾向于探究广泛得多的公共健康领域,考察自身利益和社会地位在医生、患者和国家关系中的作用。由于较传统类型工作的延续,结果就是更均衡和更多样的研究。

最大的变化很可能就是现代和当代史研究的增长。20世纪50年代,西方的科学史主要关注科学的开端以及科学革命时期现代科学的起源。做这样的研究的科学家和医生们接受的是20世纪初的教育。但是现在,多数科学史家接受的是历史系或科学史系的教育。拉丁语技能教学开始消亡,他们中很少有人懂得希腊语和拉丁语,因此他们不能对19世纪中期以前的任何时期开展研究工作。

* * *

这些演讲的目的是鼓励年轻的科学史家们思考他们能学能用的范围广泛的方法。这些演讲有五个主题:

"科学史和医学史正发生怎样的变化"讨论了以上概括的一般历史的变化如何影响了对科学和医学的研究。这一讲考察了当前在中国等地用到的一些新进路。这一讲还给出了一些应用起来会有好处的新的探究工具的例子,给出了早期中国科学的一些特征,这些特征影响了已做研究的方式。

"运用社会学和人类学方法"说明了运用这些进路的优点。二者都是作为当代社会研究的方法而产生的,但是学者们发现它们对理解从远古到现代的每个阶段的历史,都极为有用。医学中,这些进路促进了技术语言意义的认知变化,促进了对于非医生所做多种治疗的理解的认知变化。它们提供了从广泛得足以阐明古代卫生保健的方方面面去思考医学有效性的种种方式。

"运用大众文化研究方法"讨论了研究所有中国人共享文化的极有价值的方法。多数科学史研究涉及的是著名精英的工作。一些史学家试图通过讨论普通人的非常不同的进路来提供均衡,这些普通人多数直到最近才接受教育,他们不会读写。用医学作为例子,这一讲讨论了有助于这样的研究的多种多样的方法。

"比较"一讲考察了用想象处理问题时会有启发性的一类历史工作。比较对于两种不同文化的原创性理解,以及对于不同时期的同一种文化的原创性理解,都是有用的。换言之,它使得对变化进行更好的分析成为可能,这种分析毕竟是史学研究的着力点。这一讲讨论了精确比较不同语言中词语含义的重要性,列举了一些只能通过仔细比较来解决问题的例子。

"运用'文化簇'概念"一讲论述的是,要弄清那些需要多学科观点的主题的意思,就得克服历史专业化的局限性。专业化研究的基本理念,就是深刻缜密地探究问题。在19世纪的德国大学里,这个理念的创始者们相信,通才可以将这些狭隘工作的结果逐步结合起来,提供有效而综合的结果。在现代世界,这种逐步综合已经失败了。多数科学史作品是为同行专家撰写的。其他领域的学者们的理解,常常是在作品过时二十年之后。为大众读者撰写的书籍的综合性和严谨性都不足以建构一个首尾一致的知识结构。文化簇的使用允许从事原始资料研究的学者获得足以广泛的、有一般兴趣的、足以综合有效的结果。

在对古代中国的研究中,我从撰写关于自然以及人与自然的关系的作品的早期作者那里学到了许多东西。我猜想张衡、沈括、苏颂和李时珍这些人,会热心于今天的科学家们所做工作的有力结果。但同时,他们会在两个方面对现代科学极为不满。首先,早期中国人相信,这些研究的目的,在于揭示道。由于道是一个道德和美学概念,也是一个宇宙学概念,因此,对自然界的正确理解必须说明正义、美和物理实在之根由。其次,认为科学是客观的、根本没有道德和美学意义、与谈论应当如何生活无关,这样的科学观会令他们反感。他们或许比现代人更加意识到,独立于这些价值观的科学必定会导致一个环境污染的世界,会导致科学家不比没受过教育的人更在乎正义。尽管古代学者或许会尊重现代科学的精确性,但是我猜想他们会发现当代科学太狭隘。充分理解他们的种种知识观,会有助于我们拓展知识。

我读到的最启发灵感的句子出自《论语》:"知之为知之,不知为不知,是知也。"[①]多年后我终于认识到,孔子说的不是物理规律或理论之知,而是生活方式之知。

席文

2011 年 9 月 22 日

于栗山文之庐

---

① 孔子,《论语·为政》。

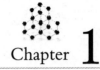

科学史和医学史正发生着怎样的变化

第一讲

20世纪50年代科学史成为职业领域时,接受科学训练的科学史学家用语文文献学工具或者考证手段分析文本,研究大人物的科学思想。到1970年,多数西方科学史学家接受史学训练,为外行写作历史。1980年代以来,不少学者把研究从古代转到当代,采取社会学和人类学方法,研究不出名的科学家、管理者甚至患者及其家人,并对社会地位、人际关系、财富、权力、说服手段和受众进行分析。近来,科学史的研究焦点已经从普适的知识体转向科学的地方性文化。北宋学人的多才多艺、现代之前科学研究的费用、科学与文学的关系、患者及其家人对于疗效的感受等,都是值得人们结合常规进路和创新进路、智识史与社会史,深入开展研究的领域。

很多人把科学史和医学史看作两个领域。那是因为60年前,写自然科学史的是科学家,写医学史的是医生。早期伟大的现代中国科学史学家之一竺可桢教授,就其所受训练和从事的职业而言,是一位气象学家。在竺教授之后的60年内开始工作的最重要的科学史家和数学史家,比如席泽宗教授和陈美东教授,受的也是科学训练。医学史家在中医学院或西医学院受过教育。医学史和科学史起初并没有很多接触,这一点我们从中国科学院自然科学史研究所与中医研究院中国医史文献研究所是分开的机构,就可以看出来。

今天,由于这两个领域彼此影响很大,其研究的问题和方法就没有很大差异了。医学史学家廖育群教授是自然科学史研究所的现任所长,这一事实就表明,这种机构的分离没有过去那么重要了。这两个研究所都把他们的研究生训练成职业的史学家。两个研究所里,就像国外大学里一样,研究的问题和方法一直都在迅速地变化。今天我将谈谈已经发生了的一些变化,以及正在持续地发生的种种变化。我将主要关注科学史和医学史发生变化的相似之处,而不是这两个领域发生变化的差异之处。出于这个原因,我将经常用"科学"一词表示二者。

职业科学史

我从20世纪50年代说起,那时科学史和医学史成了职业领域。我说"职业领域"的意思并非指高质量学术起始于那时。这两个领域的高质量学术比这要早几个世纪。人们能够作为研究者或教员谋生。1950年左右,不少学者正在从事科学史或医学史工作。他们接受的训练仍然几乎全都是科学而不是历史。他们所写的,是他们领域过去的思想如何发展成与现代科学相似的思想。在他们看来,任何不被现代标准认可的东西都不值得研究。他们用语文文献学工具,即批判地阅读和分析文本所说的内容,来批判地研究原始资料。欧洲人有他们的语文文献学传统,中国人、日本人和朝鲜人也使用类似和同样通用的证据研究(考证)工具。这两种传统对于弄清某个文献是谁写的、是否可靠、与其他文献有何联系,以及其内容的含义,是非常有价值的。

结果,科学史和医学史往往就成了概念和方法的概要,就像它们在重要书籍中按年代顺序看上去的那样。"变化"通常意味着两本书内容的不同。这种研究的目的,就是鉴别"成就"——谁最先做的、谁做得更像现在的知识。它描绘少数科学英雄,而不是普通人。好像哥白尼(Copernicus)、维萨留斯(Vesalius)、沈括(1031—1095)和郭守敬(1231—1316)与那些和他们一起工作的人几乎没有什么共同之处。这样的历史不能解释他们不同之处的原因。除了用影响之类的模糊解释之外,它们也不能说明变化。它们说不出科学家何以有时接受影响而有时却拒斥影响。为了努力理解大人物之间的关系,它们只着眼于机构①。

这种进路(approach)在长时间里成功了。其受众就像其作者一样,是科学家和医生。他们想知道谁是他们的智识先祖。他们相信他们自己时代的知识就像他们作为学生从教科书中学到的那样,是非常可靠和合理的。他们自己的工作就是改进他们时代的概念和方法。他们把历史看作一件追溯同样的改进模式的事情。他们的科学史与一般历史几乎没有多少共同之处,一般历史是关于诸种社会中无限复杂、非理性与理性相伴的人类经验的历史。

一般科学史的发展

大约从1970年以后,多数西方科学史学家和医学史学家受的就是史学教育而不是科学教育了。他们开始为外行而不是为科学家写作。他们从越南战争(1959—1975)认识到,科学技术常常被用来杀毁生命也用来拯救生命,它们可以作破坏环境的工具也可以作改善环境的工具。史学家对于科学的政治误用的关切,导致他们当中的许多人把研究题目从古代转到近代。越来越多的研究生选择研究当代科学。例如,我自己系里的12位教授中,有10位研究20世纪甚至21世纪,

---

① 黄小茹,2008,中国近现代科学史研究中的体制化刍议,《中国科技史杂志》,29(1):30-41。

我们的研究生很少研究1900年以前的事情。

研究你自己的时代时一定要注意,科学家的经济需要、他们的政治态度、他们的竞争以及他们的人际关系,都不能忽视。要注意那些接受一般训练、将永不会出名的科学家做着大部分工作,又引起了很多变化。探究所有这些事情以及许多相关的事情,正是我们今天常规研究的一部分。

这就提出了一个很显然的问题:难道同样的事情在过去就不是真的了吗?如果是真的,那么,一部主要基于大人物思想的历史就太狭窄了,不能解释科学是如何真正演进的。

大约到20世纪70年代,就在许多像我这样的科学史学家都在自问这样的问题的时候,我们发现,人类学和社会学正在发展的那些研究方法,可能非常有助于研究过去。下一讲我将处理这两个学科的影响,这里我只稍微说几句。

旧的科学史是关于英雄科学家个体的理念和理论的历史。这样的历史可以导致有价值的结论,但却不会导致均衡的结论。为了达到这种均衡,研究社会和文化在形塑技术变化中的角色,就是必要的了。今天,所有的科学家和医生都意识到社会地位、人与人之间的常规和非常规关系、财富、权力等方面的差异。这些都是社会学研究的概念。科学家和医生也从他们周围的人们那里,认识到如何理解和澄清他们的经验、如何可以使他人同意,等等。人们分享的那些感觉和方法就是我们所谓文化的东西,属于人类学。这两个学科中的多数社会科学家探究的都是当下,而不是过去。

20世纪50年代科学社会史创始时,不外是对科学机构的研究:英国皇家学会、法国科学院。由于这些机构只接纳那些通常杰出而并非典型的少数科学家,因此这种研究并不导致理解上的重要创新。从那时起,史学家把人类学和社会学用于过去,他们也就阐明了种种技术性职业的方方面面。为什么欧洲直到18世纪才有人作为物理科学家或数学家被雇用,而在过去2000年中国科学家就有了这种官职?为什么在20世纪之前的中国,科学家很少公开争论,甚至更少公开与在世的对手辩论?为什么天文官员比数学官员的争论要多一些?自古希腊人以降,科学争论和公开的面对面的辩论在欧洲很正常。这个重要反差的原因是什么?这些差异表明的是社会习俗、优先权和价值观上更深层的差异,是社会和文化上的差异。

让我通过历史地思考技术变化,给出一些重要变化的具体例子,这些例子首先是欧洲的,然后是中国的。

首先,新理念不能自动地使人确信它们是正确的而且比旧理念好。有些人不只是要发明新的技术方法,而且还要发明新的说服手段。如果这些理念是革命性的,那么,科学家还得创造要去说服的新公众。没有这些社会发明,变化可能就极其缓慢。尽管哥白尼作为一位天文学家广受尊重,但罗伯特·韦斯特曼(Robert Westman)却指出[①],在1543年他的书初版到1600年之间的三世人当中,欧洲总共只有10个人接受哥白尼的地球是行星及所有行星绕太阳运行的理论。哥白尼并非革命家。由于他是为大学里的保守学者们写作的,因此,直至伽利略(Galileo)为他创造了大学以外的新公众的时

[①] Westman, Robert S., 1980, The Astronomer's Role in the Sixteenth Century: A Preliminary Study, *History of Science*, 18: 105–147.

候,新的理解才过于缓慢地出现。

在17世纪的英格兰,大学主要是训练人做宗教职业。并非令人吃惊的是,那个时代的多数重要科学发现的完成和发展都不在大学里。由于大学并不准备赞许创新者,那么,谁来判断和接受他们的工作呢? 史蒂文·夏平(Steven Shapin)和西蒙·谢弗(Simon Shaffer)1985年对这个新问题给出了一个重要答案[1]。这就是有名望并对科学感兴趣的受过教育的绅士们,他们相互论证他们的发现和假说。起初他们只在自己家中向少数访客做论证。到1660年,他们组织了欧洲第一个科学学会即皇家学会[2]。在这个组织里,他们可以向大得多的绅士团体展示和讨论他们的新工作,并把它作为真正的科学发表在学会的学报上。这在今天,比如在《自然科学史研究》这种学报中,已经是一种正常的模式,但是在17世纪60年代它却是一项新发明。

另一个例子是标准果蝇(*Drosophila melanogaster*)的发明[3]。20世纪早期,赫尔曼·缪勒(Herman Muller)及其同事在哥伦比亚大学生物学实验室里,开始繁殖一种极便于做遗传学问题实验的特殊品种的苍蝇。繁殖了一代以上之后,他们把这些昆虫送给许多正在做同样工作的实验室。结果,这

---

[1] Shapin, Steven, & Simon Schaffer, 1985, *Leviathan and the Air-pump: Hobbes, Boyle, and the Experimental Life*, Princeton: Princeton University Press. [(美)史蒂文·谢平、赛门·夏佛,2006,《利维坦与空气泵浦:霍布斯、波以耳与实验生活》,蔡佩君译,台北:行人出版社。(美)史蒂文·夏平、西蒙·谢弗,2008,《利维坦与空气泵:霍布斯、玻意耳与实验生活》,蔡佩君译,上海:上海人民出版社。]

[2] 全称是"伦敦改善自然知识皇家学会"。

[3] Kohler, Robert E., 1994, *Lords of the Fly: Drosophila Genetics and the Experimental Life*, Chicago: Chicago University Press.

些特殊的果蝇成了多种研究的标准昆虫。这些实验室依靠它们,并通过向哥伦比亚大学的这个小组告知其工作进展及经常接受他们的建议,作为对这些赠品的回报。结果,哥伦比亚的这个小组存在的时间很长,在那种实验中取得优势地位。这样,一种昆虫成了实验科学中一种占统治地位的工具。

在这两个例子中,我们都可以看出,科学的理念和理论都依赖社会活动——新科学家之间新关系的创造,甚至一种新昆虫的创造——都使它们在某些方向的发展更可能。这也使得回答这个问题成为可能,即为什么1280年元初中国雇用了一百五十多位专家改历,但是在此后的几个世纪,欧洲却没有这样的项目把五六位天文学家弄到一起。答案在于中国的集权官僚政体。

医学史方面,过去二十多年也有同样重要的转变。此前几乎所有的研究都是关于医生及其行医的研究。医生们的著作在20世纪80年代仍然是医学史的主要原始资料,但这些资料并没有详细记载病人的经验。不过在传记和日记里,有许多这样的记载。自那时以来,我们已经广泛研读了论述患者经验的书籍,并且正开始理解普通生活中、病患中和手术中的患者疼痛史[①]。

中国医书几乎也没有给出有关患者及其经验的知识。不

---

① Porter, Dorothy, & Roy Porter, 1989, *Patient's Progress: Doctors and Doctoring in Eighteenth-century England*. Cambridge, UK: Polity Press; Porter, Roy, & Dorothy Porter, 1988, *In Sickness and in Health: the British Experience*, 1650 - 1850. London: Fourth Estate.

过,10年前张哲嘉的博士论文①(Chang,1998)使用了北京故宫档案馆的文献,非常详尽地追溯了同治皇帝(1861—1875)和慈禧太后(1874—1908年在位)在他们被医官和非官方医师治疗期间的个人看法。他们不是典型的患者,但张哲嘉的工作开辟了对一般病人进行研究的道路。

焦点变化

过去,人们都认为,现代科学是一种知识体,在哪里都是一样的。由于设想其概念和语言是普适的,学者们假定来自不同文化的科学家会用同样的方式思考和行动。但是,两个国家的化学家的心智习惯真是一样的吗?一项新近的研究提出,把科学作品从一种语言翻译成另一种语言,会改变种种意思②。作者得出结论说,这在早期是真的,直到今天也是真的。对不同文化的科学家比如中国科学家和日本科学家就相同主题所写的技术性论文,就内容和思想上的差异去做这样的研究,会是有价值的。

而且,单单研究科学概念过于狭隘,不能回答广泛的问题。沙伦·特拉维克(Sharon Traweek)做了一项值得注意的

---

① Chang, Che-chia (张哲嘉), 1998, *The Therapeutic Tug of War. The Imperial Physician-Patient Relationship in the Era of Empress Dowager Cixi* (1874 - 1908), Ph. D. dissertation, Asian and Middle Eastern Studies, University of Pennsylvania.

② Montgomery, Scott L., 2000, *Science in Translation: Movements of Knowledge through Cultures and Time*. Chicago: University of Chicago Press.

研究①,她比较了东京大学和斯坦福大学的直线加速器以及使用它们的科学家。组织在多个研究团队里的数百名物理学家需要使用这些机器。那些掌管机器的人给每项实验分配了时段。任何不能得到足够时间的项目都不会成功。特拉维克发现,斯坦福和东京决定分配时间的方式根本不同。差异并非取决于物理学,而是取决于管理实践、工作习惯以及每个地方的人际关系。一项对于日本和美国的火箭项目的比较研究,给这个结论提供了另外的证据②。换言之,要理解科学研究项目的成败,就得承认有现代科学的地方文化。

对于另外一个问题的研究还没有很大的进展,关于中国的科学文化的研究尤其如此。在每个热切专注于科学研究的国家,剽窃都是一个问题。大量研究调查了近几年美国的许多案例③。在中国,一项发表的调查提出,应当对剽窃进行历史研究④,但是再过了十几年之后,我们就不知道更多的事情了。据我们所知,20世纪之前模仿在欧洲是一种被广泛认可的惯例。但是没人研究中国的模仿。一些学者也许会把剽窃看作可耻的,但在很早的时代,它在科学上完全是件很平常的

---

① Traweek, Sharon, 1988, *Beamtimes and Lifetimes: the World of High Energy Physicists*, Cambridge, Mass: Harvard University Press. [(美)沙伦·特拉维克,2003,《物理与人理:对高能物理学家社区的人类学考察》,刘珺珺、张大川等译,上海:上海教育出版社。]

② Sato, Yasushi, 2005, *Local Engineering in the Early American and Japanese Space Programs: Human Qualities in Grand System Building*. Ph. D. dissertation, History and Sociology of Science, University of Pennsylvania.

③ Judson, Horace Freeland, 2004, *The Great Betrayal: Fraud in Science*, Chapter 7, New York: Harcourt.

④ 李佩珊和薛攀皋,1996,是英文问题,还是科学道德问题?,《自然辩证法通讯》,(4):74-80; Li Xiguang and Xiong Lei, 1996, Chinese Researchers Debate Rash of Plagiarism Cases, *Science* (New Series), 274 (5286): 337-338。

事情,如果我们想把科学理解为一种历史现象,必须对之做无偏见的探讨。

对中国科学史的新探讨

亚洲研究中国科学的史学家已经开始把许多这样的新方法和新进路用于他们自己的工作了。

日本学者山田庆儿和栗山茂久①在关于医学和科学的书中已经把智识史和社会史结合起来了。例如,山田庆儿对于授时历的研究②,就是第一部考察中国一次改历的社会、政治和建制方面的著作,该书同时也是技术性方面的著作。当我就同一次改历撰写自己的书③时,我发现这部著作极其有用。马伯英的《中国医学文化史》④,对其主题有多维度的丰富调查。冯贤亮⑤是一直在用新方法研究环境史的几位中国学者中的一位。当年轻学者接受训练从种种研究技巧中选择工具时,中国将在常规进路与创新进路之间得到均衡的发展。

---

① Kuriyama, Shigehisa, 1999, *The Expressiveness of the Body and the Divergence of Greek and Chinese Medicine*, New York: Zone Books. [(日)栗山茂久,2001,《身体的语言:从中西文化看身体之谜》,陈信宏译,台北:究竟出版社。(日)栗山茂久,2009,《身体的语言:古希腊医学和中医之比较》,陈信宏和张轩辞译,上海:上海书店出版社。]

② 山田慶儿,1980,《授時暦の道:中国中世の科学と国家》,東京:みすず書房。

③ Sivin, Nathan, 2008, *Granting the Seasons: The Chinese Astronomical Reform of 1280, With a Study of its Many Dimensions and a Translation of its Records* (授时历丛考, *Sources and Studies in the History of Mathematics and Physical Sciences*), Secaucus, NJ: Springer.

④ 马伯英,1994,《中国医学文化史》,上海:上海人民出版社。

⑤ 冯贤亮,2002,《明清江南地区的环境变动与社会控制》,上海:上海人民出版社。

使用新工具

让我列举我们知之甚少或者并不知道的一些中国科学史领域,以及研究它们将会用得上的工具。

● 1973年,在一部沈括传记中,我注意到北宋时期的许多人都异常地擅长于那个时代几乎所有的艺术和科学,从绘画和诗歌到制地图、发明、数学、天文学和炼丹术。这些学者在此前后的数百年间都罕见。30年后,还没有人解释这种有趣的现象。无疑,很多学者对此都有见解,但这是一个研究问题,而研究还没有做。确实,这个问题,即智识广度的模式,应该让具有广泛研究兴趣的人去探讨。

● 对于古代科学研究的费用,几乎就没有什么研究。对于大规模的医学贸易,随便是帝国内部还是国际的,也没有多少研究。经济史正好提供了处理这些问题的技巧。例如,清华大学李伯重教授把经济学技巧用于农业史和工业化史[1]。由对老中药铺的研究,我们知道,有丰富的记载可供定量经济学的学者们进行分析[2]。董煜宇对于北宋政府的历日专卖[3]的

---

[1] 李伯重,2000,《江南的早期工业化:1550-1850》,北京:社会科学文献出版社;Li Bozhong, 1998, *Agricultural Development in the Yangzi Delta*, 1620-1850 (*Studies on the Chinese Economy*), New York: St. Martin's Press; Houndmills: The Macmillan Presss Ltd.;李伯重,2007,《江南农业的发展,1620-1850》(社会经济观念史丛书),王湘云译,上海:上海古籍出版社。

[2] 刘永成和赫治清,1983,万全堂的由来与发展,《中国社会经济史研究》,(1):1-15;唐廷献,2001,《中国药业史》,北京:中医科技出版社。

[3] 董煜宇,2007,从文化整体概念审视宋代的天文学——以宋代的历日专卖为个案,载孙小淳和曾雄生(编),《宋代国家文化中的科学》,北京:中国科学技术出版社,第50-63页。

新近研究,提供了一个模式。

● 对于欧洲的科学和文学相关的研究在近 30 年得到了相当的发展,并且提供了许多富有成效的研究课题。这种工作在中国几乎还没有开始,尽管中国古代科学家和医生在诗歌和其他艺术方面的技能通常高于西方的同类人。文学和诗歌集成往往幸存下来,而且篇幅通常很大。喜欢文学的人都知道,诗歌表达感情,这种感情用文章很难写出来。对此进行的研究,提供了途径去接近那些会被学术所忽略的思想和感情,这样的研究当然还将阐明古代中国文学与科学之间的重要联系。

● 医学史学者最忽视的领域,很可能就是疗效,不仅是药效,还有其他治疗方法的效果。史学家倾向于要么认可早期医生成功治愈患者的陈述,要么拒斥多数陈述。两者都不是理性进路。我们如何评价疗效声称?从狭义的技术性的现代观点看,试验测试并没有给出答案。

我们从医学人类学知道,现代实验室里的疗法产生的结果,不同于原始的社会和环境里用同样的药物或同样的操作所产生的结果。要理解结果,除了技巧之外,我们还得理解那些社会和文化环境。我们也需要研究治疗者与患者之间的真实关系、他们相互作用的特征、家人或其他在场者的角色,等等。我目前的研究项目是处理一千年前的功效问题,因为一项对于疗法的各个维度的研究是理解古代卫生保健如何发展的唯一可靠的方法。这样的研究是许多学者可以作贡献的。

下 一 步

在 7 年前出版的一本书里,我的同事罗维(Geoffrey Lloyd)和我论证说,为了得出最可靠的结论,研究问题的各个维度在比较研究中常常是有用的①。比较实际上是一种广泛的探究方法。不必要求比较两种不同的文明。我们也可以探究同一个社会的两个时代或两个地方。

史学家趋向于仅仅研究思想史或者仅仅研究社会史,但是这些倾向都是基于欧洲习惯的区分。在研究中国古代科学家时,还必须理解他们中的多数人所属的官僚文化,以及科学家彼此说服的方式,等等。我提出,最有效的进路就是首先要问哪个维度——个人的、政治的、社会的、经济的、组织的、艺术的、数学的等等——与给定的问题相干。然后,我们可以考察那些相关的各个维度。这让我们不仅理解了与问题相关的每个维度,而且还理解了它们是如何相互作用的。在这个系列演讲中的后续演讲中,我将论述比较和文化簇(cultural manifold)。

我最后想留给你们的,是可能相当大地扩展研究中国科学的学术范围的理念。竺可桢教授 65 年前写的一篇文章用这些话结尾:

---

① Geoffrey, Lloyd and Nathan Sivin, 2002, *The Way and the Word: Science and Medicine in Early China and Greece*, New Haven and London: Yale University Press.

苟能引起博雅君子对于本问题之探讨,则此文为不虚作矣。①

---

① 竺可桢,1979,二十八宿之时代与地点,载《竺可桢文集》,北京:科学出版社,第253页。

Chapter 2

第二讲

运用社会学和人类学方法

  人类学家的目标是发现一个民族甚或一个群体共享的、不同于他者的意义结构。共享的实在意义的全部组成,称为文化。有很多方式把所有的身体不正常划分成为种种疾病,不同文化中的人则做出不同的选择。因此,两种文化中的人不会患完全相同的疾病。信念可以引起和战胜真正的肉体疾病,因为身体对信念和意义做出响应。社会学家的目标是理解个体生活于其中的社会的结构、规范和价值观。社会学可以帮助我们理解不同社会里医生权力的差异,权力是他们治愈疾病的能力的一个重要因素。新的疾病和疾病名称的改变,并非仅仅基于研究上的改进。医生就像其他社会成员一样,对其社会广为流传的价值观做出响应。如果他们不做出响应,他们就不可能非常成功地治愈患病的人。

第二讲 运用社会学和人类学方法

我的话题是人类学和社会学,大部分例子来自医学史。这是因为医学社会学和医学人类学发展得不错,提供了不少好范例。社会科学对科学技术研究(studies of science and technology)也是格外有用的。

首先,我假设大家没有在这些领域学习过。我将界定这些领域,并解释它们在我们所研究的各类历史中的作用。这两个领域都古老,19世纪就开始呈现出现代形态。起初,这两个领域在目标和方法上明显地不同。

20世纪的著名人类学研究,多数涉及的不是西方民族,通常都在欧洲势力范围的殖民地中,比如非洲人和南洋人。研究的目标是对文化的研究,而"文化"则是一个有多种含义的词。在上个世纪的人类学中,它开始表示一个民族所共享的、并用以对其经验进行归类和理解的模式。例如,在中国文化里人们把褐色当作深黄色,而在美国文化里褐色却与黄色无关。两种文化都没错;这不过是一种文化差异。

另一方面,社会学始于学者分析他们自己的社会结构。他们探究社会地位的差异、财富的差异、权力的差异,等等。这些差异多因国而异。例如,在美国和中国,教授和学生之间的关系就大为不同。两人彼此相处的方式取决于这些差异。

我将首先讨论人类学。

人 类 学

人类学有若干分支。我今天将谈论的一个分支，在美国通常称为文化人类学，而在英国则称为社会人类学。① 这门学问在发展出所谓参与者观察的技巧时，就变得重要起来。人类学家生活在与他们不同的人群之中，参与他们的工作并观察他们。人类学家研究他们的日常生活以及他们特殊的定期仪式（礼仪、礼俗）。其目标是发现一个民族甚或一个群体共享的、不同于他者的意义结构。人们如何看待自己与他人？人们如何相互作用？人们如何理解自己的社会与非人类世界，如何对之归类？人们如何理解变化？他们对诸如健康、疾病和康复之类的事情如何做出解释？这些就是共享的实在意义的全部组成，人类学家称之为文化。多数人类学家的目标就是理解文化模式。②

用汉语谈论仪式是困难的。因为研究人类学的人并不多，许多学生都把仪式看作非科学的，因此不值得研究。但是人际关系并不是科学的，如果我们想理解它们，我们就要使用一些允许我们分析其复杂性的概念。人类学和社会学提供的

---

① 这两个专业的区别不少，可是在这里并不重要。这次演讲不是要对各门社会科学做一般的介绍，而仅仅是概述我发现的在科学史研究中特别有用的两个方面而已。

② 人类学是一门多姿多彩的学科，不可能找到适合初学者阅读的一本单一的书描述它。要理解其当下——总是处于变化之中的——状态的意义，我推荐阅读如下期刊最近各期：《美国人类学家》(American Anthropologist)、《人类学与医学》(Anthropology & Medicine)、《皇家人类学研究所所刊》(Journal of the Royal Anthropological Institute)和《医学人类学季刊》(Medical Anthropology Quarterly)。

许多极为精妙的概念和方法，证明有助于对许多文化的研究。

仪式观念就是这些概念和方法中的一个。这个概念比中国古代的礼仪观念要宽泛得多。它指的是所有有效的被模仿的行为，而且还有超越行为有效性的意义。这个房间的每个人都理解，今天主持人介绍我之后，人们期待着我要演讲，期待着各位要听。主持人的介绍有效地提供了我的信息，但是其意义也是让各位有个准备，听我的演讲而不是听各位说话。每位学生都学会了理解这种介绍仪式。这就是它有效的原由，就是其意义赋予演讲者以一定权威的缘由。平常的生活充满着仪式。它们呈现出基本的文化模式。

让我举一个仪式有效性的例子。一个人在生活中从一个身份转变为另一个身份，就是人类学中许多有趣的话题之一。多数社会里，主要的转变就是诞生、从童年变为成年、从单身到结婚成家，以及去世。对于其中的每种变化来说，通常都有一个仪式或典礼，指引人们从一个旧身份转变为一个新身份：生育典礼、成人典礼、婚礼、孩子出生典礼，以及死亡典礼。婚庆用象征符号和特殊行为教少男少女如何成为夫妻，为他人接受他们成为夫妻有个准备，表示已经合法婚配，等等。

在中国，古代统治阶级复杂详尽的典礼多记载在《礼记》里。今天在中国，城乡之间、社会层级之间、地区之间、富者与贫者之间、民族之间，这些转变的典礼都不同。但是每个人都经历同样的生命阶段，其亚文化都在教他们庆祝和哀悼的方式。

学者们建立他们自己的小文化。一位出席中国科学技术史学会会议的人类学家，可以通过了解官员们做的是什么事情、谁在会上发言、如何选择论文、谁决定每篇报告的论文、谁在小组里最先发言、人们在过道里如何相互指责等等,来分析那个小文化。

一位人类学家想了解物理学文化,可以在某人获得博士学位时、某人在研究所或学系开始工作时、某人成为物理学会的官员时、某人退休时等等,来看人们的举止[①]。所有这些小型的入会礼和庆典,就像家族的礼仪一样,都是仪式。仪式把研究生转变成为研究助理,并赋予其研究者同事一个正规方式接纳他为合作者。仪式可大可小,大如大学毕业典礼,小如研究所领导介绍一位新同事。[②]

人类学与医学

一旦认识到文化型塑了人的所有选择和行动,我们就能明白文化是理解医学的关键——医学在不同地方如何不同,怎样随时间变化。医学人类学家在很久以前就知道,两种文化中的人不会患完全相同的疾病。有很多方式把所有的身体不正常划分成为种种疾病,不同文化中的人则做出不同的

[①] Peter Galison, *How Experiments End*, Chicago: University of Chicago Press, 1987. 这是关于这些问题和相关问题的一项著名的研究。

[②] 参见:Bruno Latour, 1987, *Science in Action: How to Follow Scientists and Engineers through Society*, Milton Keynes: Open University Press, 1987. (布鲁诺·拉图尔著,刘文旋、郑开译,《科学在行动:怎样在社会中跟随科学家和工程师》,北京:东方出版社,2005年。) Bruno Latour & Steve Woolgar, *Laboratory Life: The Construction of Scientific Facts*, Princeton: Princeton University Press, 1986. (布鲁诺·拉图尔、史蒂夫·伍尔加著,张伯霖、刁小英译,《实验室生活:科学事实的建构过程》,北京:东方出版社,2004年。)

选择。

如果我们假定任何地方的患病经验都一样,我们就会犯严重的历史错误。例如,人们阅读诸如《伤寒论》之类的旧书,通常就认为"热"是"发热"的意思。在生物医学上,"热"是指体温高于正常值。但是在早期的中国作品中,"热"通常不是指医生测量出的体表温度,而是指患者经验并告诉医生的体内热感。① 它与体内冷感"寒"相对。

《伤寒论》甚至19世纪的传统医书中,"伤寒"本身通常并非伤寒——现代词典给出的仅有的意思——而是涉及不同发热种类的许多类不同疾病。这就是我把"伤寒"按词义译为"Cold Damage Disorders"(冷损害紊乱)的原因,为的是提醒读者注意,古代之中国的疾病经验与今天是不同的②。

人们在成长过程中知道了什么是病患什么是健康。他们从双亲那里学会如何做病人——就是说,他们感觉不舒服时人们期待他们有怎样的表现、求谁帮助,等等。帝王时代,医生对疾病分类的方式发生过很大变化。自西医引进以来,疾病分类发生了更迅速的变化。

我们一想到医学就会想到医生。但是人类学家提醒我们,处理病患的人是广为不同的。卫生保健的第一步是自疗。

---

① "热"始终在这种意义下使用。《伤寒论》里最早出现过"发热"一词,但多数情况下,其用法太模糊,不能确定它指代的那个阶段的热感和高体温多少时候一次。

② 我给出了许多关于这类差异的例子[Nathan Sivin, Traditional Medicine in Contemporary China. A Partial Translation of Revised Outline of Chinese Medicine (1972) with an Introductory Study on Change in Present-and Early Medicine, *Science, Medicine and Technology in East Asia*, 1987, (2), Ann Arbor: University of Michigan, Center for Chinese Studies.]关于单一中文疾病名称的意义变化的详尽研究,见 Hilary Smith, Foot Qi: History of a Chinese Medical Disorder, Ph. D. Dissertation, History and Sociology of Science, University of Pennsylvania, 2008。

如果你早晨醒来时头疼,你不会直接去看医生,而可能仅仅服用阿斯匹林或是正天丸。如果你不确定自己是怎么了,你会去问你的母亲、妻子、友人或者其他你信赖的人的意见。这通常是卫生保健的第二层次。如果你仍然感到没有好转,你或许会去药店要人推荐药。如果这些层次的卫生保健都失败了,你可能去诊所。卫生保健通常包括许多层次,只有最高层次才有职业医生。

传统中国也是如此。事实上,多数人从不看医生。中国人太多,多数人没什么钱,而医生又太少。在乡村,一位农妇很可能会把她患病的孩子带去找一位在山里采草药的邻居,或者去找一位铃医[①]或法师。对于一位史学家来说,理解其治疗价值并不容易——但理解一位医生的疗法价值也不容易。今天,任何未经实验室试验证实的关于医学成就的报道,都被受过教育的人们所拒斥。但是,医学人类学在很早以前就表明,这个标准过窄就没有用。

多年前,当时我和妻子住在英国的剑桥。一个夏晨,我醒了,站起来,立即就跌倒了。这看似很奇怪,我站起来,又跌倒了。我不再能控制自己的身体。我问自己是怎么回事。如果是头痛,我应该会意识到,并会服用阿司匹林。这种情况下,我不知道是怎么回事。唯一的想法就是:"如果长期如此,那将多么麻烦!"不久,我妻子注意到我躺在地板上。下一步的确就是询问她的意见。结果,她给一位友人打电话,很快就弄明白了,一种奇怪的新病毒也让几位友人都患上了感冒,等 48

[①] 铃医亦称"走乡医"、"串医"或"走乡药郎",指游走江湖的民间医生。铃医以摇铃招徕患者,故而得名。——译者注

小时就会好。由于我们不再担忧,我也就不必叫医生了。一天之后,症状开始逐渐消失,我确信不必治疗,就那样得了。

让我们想象古中国的一个类似情况。某人早晨不能站起来,就像我一样感到不再能控制自己的身体,很无助,但其感受严格地讲与我的不一样。他会听说一系列可能的原因去思考,这些原因与我所知道的不一样。从前人们会考虑——现在中国部分地区仍然会考虑——的一个原因就是,鬼已附身。其妻子会请来一位法师把鬼驱赶出去。尽管今天看来很奇怪,但人们相信法师有控制鬼神的权威。不论这种控制是客观事实还是无关紧要,反正人们信它。这种信念往往比药物更有效。法师的仪式设计得就像我妻子打的电话一样,使患者确信其生命并没有真正失控,预期不久就会恢复常态。信念的确可以战胜真正的肉体疾病,因为信念可以引起真正的肉体疾病。身体对信念和意义做出响应。这是医学人类学最伟大的一个发现。

中国和西方的医生也都用信念治病。医生的行为方式、他所使用的语言、他使用仪器的方式,都引起患者发生变化。仪器不仅仅是诊断和治疗的工具。它们也是医生的知识和技能的象征,医生能用而他人不能用的象征。它们是包括医生和患者在内的某种仪式的一部分,我们可以把这种仪式称作一种科学仪式。用科学的能力受人敬佩,给医生以极大权威以序治乱。这就是人类学家认为我们要想充分理解现代科学就需要分析其技术价值和象征价值的原因。

事实上,治愈疾病有三种情况。

首先,多数患病的人即使不接受治疗也能恢复健康。人体在能响应反常时,就直接响应反常,恢复正常。希波克拉底(Hippocrates,公元前460—公元前370)以降,医生们就已认识到,他们最有效的方法不过是支持患者自我恢复。这也是《黄帝内经》和其他中国经典中的一个重要主旨。对于多数疾病,所有医生需要做——或者能够做——的一切,就是促进恢复和处理症状。我们可以把这种恢复倾向称为身体的自响应。

第二,中西医都基于的假设,就是物理和化学治疗能够影响人体过程并战胜疾病。我们可以将此称为技术响应。

第三是身体对仪式和其他意义象征的响应。有些人类学家将此称为意义响应。所有这三种响应都可能同时发生。

并非很多人类学家都研究过去,不过,那样的洞见对历史研究同样有价值。总之,对疗法的充分分析包括身体自愈能力、疗法的仪式和象征环境的效应,以及技术方法的价值。[1] 与一位只懂技术部分的医生相比,一位受了训练懂得疗法三要素的医生可能是更有效的治疗专家。对于研究过去的医学史学家来说,同样的理解也是有益的。类似地,在天文学史方面,知道当时的人们怎么思考天空和事件,与懂得预言技能同样重要。例如,孙小淳和一位同事就撰写了一部论述汉代的人们如何看待天空的很有价值的著作。[2]

---

[1] 对于医学效能的这三个组成部分的最佳讨论,见 Daniel Moerman, *Meaning, Medicine and the "Placebo Effect"*, Cambridge Studies in Medicine Anthropology, Cambridge and New York: Cambridge University Press. 2002。

[2] Sun Xiaochun and Jacob Kistemaker, *The Chinese Sky during the Han: Constellating Stars & Society*, Leiden: E. J. Brill, 1997.

社 会 学

社会学的基本理念是,为了理解个体的活动,我们必须理解个体生活于其中的社会的结构、规范和价值观。每个社会对于正常的人际关系是什么、人们如何举止、他们应当视什么为好坏,都有自己的理解。个体表现并非总是严格符合他人的期待,人们也不抱有如此严格的期待。这些结构、规范和价值观都是理想,在教人们如何生活方面很重要。父母把它们教给孩子,老师把它们教给学生,甚至电视也把它们教给观众。这就是我们认识到中国人按某些方式行事,美国人按另外的方式行事的原因。我们可以明白,华侨们的行事方式既反映了其父母教的东西,又反映了他们从一般美国人那里学到的东西。今天,我将讨论两个社会学话题,即职业概念和疾病分类学的变迁。

中文"职业"一词既表示"occupation"又表示"profession",但在西方关于社会的思想里,其意义完全不同。这是因为数世纪的文化差异。按照社会学家的看法,任何职业都是occupation,但是只有需要水平特别高的教育并且由此获得高的社会地位的职业才是profession。

在早期欧洲大学里(大约1200年以降),只有三个群体接受的教育在基础教育以上:神父或牧师、律师和医生。不像其他职业,这些群体自身就赋予其中的人们三种特殊的权力:这些群体自身控制着谁从事这些职业、决定付给成员多少钱,而且可以阻止无资格者从事这些职业。19世纪以来,神父和牧师失去了很多社会权力。今天,医生和律师是两个标准职业。其他群体,比如科学家,试图获得同样高的地位。他们使用同样的方法——研究生教育等等。但是多数科学家都是大学和其他机构的雇员。他们不能分配自己的收入,也不能控制从事科学的人。

医学社会学

在多数欧洲国家和美国,医生和律师仍然是所有职业中有最高威望者。然而,遍布许多国家的公营或私营医疗保险的特殊体制,已经夺走了医生对自己收入的控制,使平均收入少多了。越来越多的律师成为雇员,而不是作为其他律师的伙伴而工作。因此,这两种职业的自治权和职业特征正在消失。在某些国家,这种特征就消失了,尤其是在社会主义国家,这些人是政府雇员。

由于中国社会的特殊特征,医学是一种 occupation,而不是从历史意义上的 profession。在古代,任何人都可以行医而不管其受到何种教育,多数医生都是老医生的弟子,而不是在特殊学校里接受训练。他们不需要专门的学位或者行医执照。政府举行医学考试长达许多世纪,不过这些考试通常只是针对太医局官职的,并不是针对一般行医的。太医局成员的确是官员,但是其职位并不高。①

没有官职的医生常常抱怨他们称为庸医的那些人的竞争,但却不能阻止他们行医。医生的收入取决于其个人声望,一定程度上取决于其老师的声望。但医生的声望不一定取决于疗法。许多医学家有名,是因为他们的非医学高阶官职或者是因为他们是杰出的文学家。这样的人给友人治疗,并不打算靠医术谋生。宋代两个很著名的例子是沈括(1031—1095)和苏轼(即苏东坡,1037—1101)。但是,那些作为医生来谋生的人拥有许多不同的社会地位,他们不得不与各种宗教治疗师及大众治疗师竞争。20世纪20年代之前,医生并没有组织成一个职业。

新中国成立以来,是政府来决定谁是医生、分配收入,决定谁不再有资格行医。许多全职或兼职私下行医的医生可以增加收入,但是政府仍然控制着大部分行医活动。

总之,世界各地的医生职业都不同。他们的社会地位因国而异,并且在每个社会都发生了变化。我关于职业和工作的观点就是,社会学可以帮助我们理解不同社会里医生权力的差异。我们已看到,权力是他们治愈疾病的能力的一个重要因素。

疾病社会学

社会学里的另一个重要问题就是如何定义疾病。从前,

---

① 详见梁峻,《中国古代医政史略》,呼和浩特:内蒙古人民出版社,1995年。

人类学家就发现,不同的文化有不同的疾病。并非唯有科学研究才能决定什么是一种疾病。即使是在接壤的欧洲国家,医生对疾病和疗法的理解也相当不同。在任何一个国家,对什么是一种疾病的理解都在持续而迅速地发生变化。① 让我举两个美国的例子,在这两种情况下欧洲的现状则不同。

- 18世纪时基督教在美国有极大的权力,常醉酗酒是一种违背宗教伦理的罪孽(sin),被神父或牧师指责是有罪的。19世纪和20世纪初,常醉酗酒成为非法,是受法官惩罚的法律上的罪行(crime)。当我还是小孩的时候,当众醉酒的人常常被捕。20世纪末,酗酒成为一种疾病,它的名称变成了"酒精中毒"。

这并不是因为医生发现了一种疗法。他们又不能解释酒精中毒的原因。社会学家论证说,醉酒从一种宗教道德上的罪孽转变为一种法律上的罪行,是宗教权力的缩减和国家权力的增长的结果。醉酒转变为酒精中毒,或许是由于对醉酒的惩罚过多无效所致。"醉酒"暗示某人是坏人;"酒精中毒"则是指一种病患,而并非个人过错问题。医学无能力治愈它无关紧要。医学不能治愈许多其他严重的身体问题,比如癌症。尽管医学不能治愈酒精中毒,但是很多人仍希望有一天医学在法律失败的地方取得成功。

- 在美国,老衰或老年痴呆过去相当常见。得这种病,老年人失去记忆,变糊涂,甚至经常连家人都不认识。同时,阿

---

① 佩耶巧妙地研究了英、法、美、德的这个问题。尽管在最近20年里,在这四个国家都发生了很大变化,但至今没有类似的研究。见 Lynn Payer, *Medicine & Culture: Varieties of Treatment in the United States, England, West Germany, and France*, New York: Henry Holt, 1988, 1995。

耳茨海默氏病（Alzheimer's disease）是一种罕见的疾病，患此病的青年人有相同的问题。二者症候几乎相同；主要差别是年轻人极少出现这些症候。两种病都不能治愈。奇怪的是，过去20年里，医生已经停止诊断老衰。现在，即使患者很老，他们都用"阿耳茨海默氏病"的说法。这已经从罕见的诊断变为常见的诊断，而且医生不再判定这种病是老衰。这种变化并非来自新的科学研究。更确切地说，这似乎是人们认为一个不好的词变成了一个中性的词。

很容易找到同类的例子。① 这些例子暗示，新的疾病和疾病名称的改变，并非仅仅基于研究上的改进。医生就像其他社会成员一样，对其社会广为流传的价值观做出响应。如果他们不做出响应，他们就不可能很成功地治愈患病的人们。

如果一位医生认为过量饮酒并非坏行为而且也不是病，那么，他在酒精中毒的治疗中就扮演不了什么角色。如果其他医生在谈论阿耳茨海默氏病时，他却坚持称患者是老衰，那么错的就是他而不是他人了。类似这样的例子表明，疾病的命名是一个社会现象，在这种现象中医生和外行彼此影响直至他们达成一致。事实上，某种特定的诊断有时成为一种时尚。这是社会学有必要据以充分理解医学的诸多方式中的一种。

---

① 见阿茹诺维茨对疾病变化所做的许多这样的分析，Robert A. Aronowitz, *Making Sense of Illness: Science, Society, and Disease*, Cambridge History of Medicine, Cambridge and New York: Cambridge University Press, 1998. 亦见 Arthur Kleinman, *Social Origins of Distress and Disease: Depression, Neurasthenia, and Pain in Modern China*, New Haven: Yale University Press, 1986（凯博文，《苦痛和疾病的社会根源：现代中国的抑郁、神经衰弱和病痛》，郭金华译，上海：上海三联书店，2008年）。

结　　语

人类学家和社会学家有着不同的出身,发展了不同的概念。但是在过去一百年间,他们阅读彼此的著作,意识到许多相似之处。一些人类学家终于开始研究精妙得像他们自己的文化那样的文化——例如中国文化。我的一位老师过去常常说,如果中国人类学家对美国社会进行研究,那么,美国人就能学到很多东西。现在,有许多人类学家研究他们自己的文化。[①]

同时,社会学家发现,他们的技巧在研究不同于他们自己的社会时很有用。自然,两个群体终于学习既使用人类学的文化概念又使用社会学的社会概念了。只是学科建制并没有迅速变化。两者仍然有不同的专业协会、学报等等。

我的下次演讲论述通俗文化,将着手处理人类学和社会学在科学史和医学史中的其他用途。

---

① 第一位研究中国文化的中国人类学家是许烺光,见 Francis L. K. Hsu, *Religion, Science and Human Crises: A Study of China in Transition and its Implications for the West*, International Library of Sociology and Social Reconstruction, London: Routledge & K. Paul, 1952。

Chapter 3

第三讲

运用大众文化研究方法

　　大众文化是医疗保健、农业、制陶、金属工具制造等许多行业的基础。批判地利用宗教作品、医书、史书、笔记、文学作品等许多文献,可以拼合出大众文化中的科学、技术和医疗活动图景。中国古代医生的学识使其成为宇宙动力学专家,大众法师的灵魂知识赋予他在神官系统的影响。《诸病源候论》讨论的"产后血运闷"疾病,《千金翼方》记载的"产运"疗法,表明禁忌和仪式治疗在某种程度上是有效的。在大众文化研究中,人类学和社会学提供了经受过检验和验证的工具。

我们所了解的中国科学和医学,多是有关精英(即中国统治阶级)的工作。在中华帝国肇始之前,孟子就描述了统治者与其他人群之间的差异:

> 或劳心,或劳力。劳心者治人,劳力者治于人。治于人者食人,治人者食于人,天下之通义也。①

正如大家所知道的,治者与受治者之间的这个裂隙持续贯穿整个中华帝国的历史。有许多基本的差异。整个士阶层有资格成为官员,并且享有许多特权。至少到宋代,这个阶层的大部分都拥有土地并征收地租。精英与布衣之间最重要的差异在于精英受过古典教育,他们有文化。如果我们研究历史,我们就得利用他们撰写的书籍和文献。这在很大程度上意味着,我们通过他们的双眼去看他们周围的世界。例如,当孟子谈论"天下之通义"的时候,其本意并非说那些全靠劳作谋生甚至还要养家糊口的贫穷农夫接受此通义。没有证据表明,孟子要贫穷农夫明晓大义。

北宋以降,多少有些文化的商人和工匠的人数在增长。自明以降,少数商人和工匠与官员一样受教育。不过,多数普通人尤其是贫穷农夫,仍然是文盲。

---

① 《孟子·滕文公章句上》。

我们对未受教育者的了解,多半来自统治他们即他们供养的那些人的作品。史学家都认识到这些作品不客观。我们必须辨识每位作者的偏见。如果这么做,我们就能根据许多作品拼合出一幅精英之外的人们的一般生活图景来。

我今天主要谈卫生保健,因为多年来我的工作一直就是研究这个方面。不过,我用的方法对许多其他技术性论题也有用。

正如我在上次演讲①中提到的,行大众疗法者中,有受教育程度不及行古典医学(classical medicine)②的医生,有山中采草药的农民,有铃医,有打坐冥想的师傅,还有道士、佛僧,等等。他们各有不同的关于健康和疾病的理念,疗法也不相同。

我们发现医书里有很多关于普通人疾病的讨论,也有关于非医生医治的大量信息。由于这样的信息在中国人的卫生保健上起着很大的作用,我认为值得进行严密的考察。

除了医书之外,还有很多其他信息来源。由于治疗疾病在每种宗教里都很重要,我已经在宗教作品尤其是《道藏》里找到了很多材料,《佛藏》里多少也有一点。散见于包括史书和笔记在内的各种文学作品里,也有宝贵的资料。③

批判地使用这些信息,我们就可以拼合出一幅相当清晰

---

① 席文,"社会学和人类学方法之对于科学史和医学史的应用",《清华大学学报(哲学社会科学版)》,第25卷(2010)第6期,第5—10页。

② 我用"古典医学"表示20世纪之前的中医。它与现代传统中医(TCM)相当不同,受外国医学的影响很小。

③ 对于这项工作来说,有用的参考书包括:陈邦贤,《二十六史医学史料汇编》,北京,中医研究院中国医史文献研究所,1982;陶御风、朱邦贤、洪丕漠编,《历代笔记医事别录》,天津,天津科学技术出版社,1988;中医研究院医史文献研究所编,《医学史论文资料索引》,第二辑(1979—1986),北京,中国书店,1989。

的大众文化中的医学图景。由于它与精英医学大异其趣,我将首先描述精英医学的基本理念,然后再比较大众医学的理念。

精英医学的宇宙观

公元前 250 年之后的两个半世纪,很多重要的哲学家提出了包括天地、天下和人体在内的宇宙观。[①] 自然界即宇宙按照一定的周期运转。哲学家们用阴阳五行的术语描述这些周期。明暗的日循环、暖冷的年循环、日月星的运动,都是有规则和永恒的。一个国家治理得完好,其活动就遵循着与自然界所遵循的同样的节律,因此这个国家在理论上也就会永远延续下去。皇帝作为"天子"应该负责维持天地秩序与天下秩序之间的和谐——这就是良治的含义。

同样,只要保持身体与自然循环变化的和谐,身体就健康。这就要求按照正确的方式生活,避免过分的行为、思想和情感。疾病侵袭的是没有保持这种平衡的人。医生的工作就是确定什么失衡了以及如何来恢复这种平衡,其主要的技术性分析工具就是哲学家的阴阳五行概念,以及基于这些概念

---

① 最重要的是《吕氏春秋》(公元前 239 年)、《淮南子》(公元前 139 年)和《春秋繁露》(公元前 156 年及以后)。对于天地、天下和医学的综合,见《黄帝内经》(公元前一世纪?)。G. E. R. Lloyd 和 Nathan Sivin 在 *The Way and the Word: Science and Medicine in Early China and Greece*(New Haven, Yale University Press, 2002)一书的附录中有对这种综合及其形成过程的详尽分析。

的其他概念,诸如六经之类。① 他从师傅那里或者论述本草和其他疗法的书本上,学会可以用什么资源去纠正他所发现的那种失衡。

换言之,精英医生的技能在于分析宇宙和身体过程,决定如何把它们带到那种健康的和谐状态。②

大 众 医 学

大众医学包括所有种类的疗法,是大众文化的一部分。这个术语需要定义。它并非与精英文化相反,而是范畴更大。把它翻译成"普遍文化"要比译为"民间文化"更确切。③ 中国社会还没有进入当代的时候,每一个中国人都随着自己的成长而学会了一种关于个人、社会和自然的观念。每个人都熟悉它,无论他们是否受过教育、是否有权势。男性精英成员既学习这个观念,也学会了精英文化。某些情况下,比如为

---

① 每个医学院的"基础理论"教科书都解释了这些概念和其他概念及其用法。这些概念的古代意义和现代意义的差别,见 Nathan Sivin, *Traditional Medicine in Contemporary China: A Partial Translation of Revised Outline of Chinese Medicine*, Ann Arbor, University of Michigan, Center for Chinese Studies, 1987, 43-196. 此书1972年初版,1987年版加上了一篇介绍当代和早期医学研究和变化的文章,收入《东亚科学、医学与社会》第2号。

② 古代中国并没有单一表示健康的词。医书作者用诸如"平人"之类的多个词,来描述健康的人。今天,"健康"这个普通词,是现代从日本借用的。关于早期医学中的健康和失调,见 Nathan Sivin, *Traditional Medicine in Contemporary China: A Partial Translation of Revised Outline of Chinese Medicine*, Ann Arbor, University of Michigan, Center for Chinese Studies, 1987, 95-111.

③ 今天,在消费者社会里,"大众文化"主要指花钱偏好。英语"大众的"一词从15世纪开始用来指"粗俗"或下层文化,但自18世纪后期以降,"大众的"一词在学术上的主要含义就是"广为流传的"了。

官——他们轻视大众文化;也有时候——例如生活在故里时——他们就积极支持它。

因此,"大众医学"比"民间医学"这个术语更加确切。今天我要讨论的是汉人的大众医学。中国的少数民族一般有他们自己的信仰和习惯。大众医学的意思是汉人所共享的有关身体、健康、疾病和疗法的一种理解。它并非基于精英医学的宇宙进程观,而是基于对灵魂社会的理解。分析这套观念,可以解释大众医学起作用的方式。

普通人认为,与人类社会相伴的还有一个灵魂社会,我们可以把灵魂社会划分为神、鬼和祖先。某人死了,用特定仪式和供品照顾这个人的亡灵就是家人的责任。礼仪做法得当就使死人成为祖先。这样的灵魂就会对其后裔行善。鬼是没有得到适当照顾的亡灵,因此会伤害其后裔和他人。神是一种不仅对其后裔,也对某个群体、某个社区乃至所有的人行善的灵魂。① 治病是最重要的一种善行。当治愈者超过了一个家族的时候,这些治愈者就会向这个灵魂祈祷并供奉,于是,这些人就开始把它当做神了。如果对这样的灵魂的崇拜广泛传播,它就会成为普遍公认的神,而且出于这个原因,人们就会给它在天官系统里指派一个官职。天官系统是按与帝王官职系统相同的方式组织起来的,它具有类似的治理灵魂世界的权力。

不同家族的人对谁是鬼谁是神通常意见不一。我的祖先可能是你的鬼或者是我们村庙里的神。就像帝王官职系统一

---

① 并非所有的神都是死者的灵魂,有些神是人格化了的宇宙面貌。

样,没有不同种类的人,但是死人有不同的功能。

讨论大众医学的原始资料显示,多数疾病是着魔。就是说,危险的灵魂或是鬼魂侵袭了身体,干扰了其正常过程。这可能是不同原因引起的。常见的原因是不道德或劣行。忽视家庭故人的亡灵就是一个例子,其结果是污秽。健康人在精神上和道德上是纯洁的。如果一个人不纯洁,亡灵就会侵袭其身体,使他生病。

大众医治者——例如法师——是神灵机构里的专家。他们能够弄清楚是哪个鬼魂造成麻烦,知道如何结束着魔。为了理解这一点,我们琢磨一下帝制中国的衙门。多数普通人不了解它,也不跟它打交道。但是,一个来自衙门的行人或者步快①可能迟早会为某件不重要的事情找你要钱。如果没有钱,你会怎么办?如果你想去衙门告发他,你很可能会陷入麻烦。但是如果你认识村里的某个人,他在衙门里有朋友,你就会请这个人帮你。这就像大众法师的技能一样:他知道天上机构的一些事儿,知道用仪式去寻求帮助。一旦他弄清了是哪个灵魂进入你弟弟的身体,你的家人就会参加法师的仪式去除掉这种污秽,把这个灵魂赶走。

这是古代几乎每个中国人都相信的。现在还相信的也不少。全世界的人类学家的工作都表明,信念可以致病,也可以治病。运用他们的洞见,就开辟了充分理解中国卫生保健史的道路。

---

① 行人是官府的小吏差役。步快是捕快的一类。捕快负责缉捕罪犯、传唤被告和证人、调查罪证。配备马匹执行公务的捕快称为"马快"。徒步的捕快称为"步快"、"健步"或"楚足"。——译者注

现在我来总结一下我对精英医学与大众医学所做的比较。首先,它们基于对身体和环境关系的不同理解。精英医生把健康理解为人体与宇宙的动态节律之间的和谐,把疾病理解为打破这种和谐,把诊断理解为判断失调的是什么,把疗法理解为重建和谐的种种方式。大众法师把健康理解为身体在道德和精神上的纯洁,把疾病理解为污秽精神的着魔,把诊断理解为辨认迷住心窍的灵魂,把疗法理解为恢复人体纯洁的种种仪式。这种纯洁反过来恢复了个体与周围人的和谐。用另一种方式表达就是,医生的学识使他成为宇宙动力学专家,大众法师的灵魂知识赋予他在神官系统的影响。

效　力

接下来,显而易见的问题就是,医学人类学和医学社会学是如何帮助我们理解疗法的。它们关于大众医学的效力认识到了什么？我在上一次的演讲里提出,三个东西的结合治疗疾病:

- 身体自然恢复的能力：身体的自身响应。
- 药物和其他疗法引起的体内化学和物理变化的能力：技术响应。
- 身体对仪式和其他意义符号做出的响应,也能引起体内的化学和物理变化。一些人类学家称之为意义响应。

理解前两种并不难,一个例子或许有助于理解第三种。一些精神病医生研究了中国的大众疗法,而由于他们是心理学专业的,他们就得出结论说,仪式可以造成患者的心理变化。①他们论证说,仪式是一种原始的心理疗法。这在一定程度上是对的,但却是一个过于狭窄的解释,疗法不仅影响心身,而且——由于每种文化里人都知道何以患病——它也影响社会关系。我的详尽的例子将说明这一点。

一部很重要的论述病因的早期著作《诸病源候论》(公元610年),讨论了一种称之为"产后血运闷"的疾病(生孩子后眩晕和作呕)。这是很严重的疾病,母亲可以致死。这部隋朝御医写就的书,用古典医学的语言描述和解释了这种疾病,强调了阴阳血气在循环系统中的平衡:

> 运闷之状,心烦气欲绝是也。亦有去血过多,亦有下血极少,皆令运。

文本详细描述了该病的不同类型以及如何区分它们。书中指出了这种疾病何以危险:

> 然烦闷不止,则毙人。

书中接着解释了导致这些困难的原因:

---

① 最著名的是 Arthur Kleinman (ed.), *Medicine in Chinese Cultures: Comparative Studies of Health Care in Chinese and Other Societies: Papers and Discussions from a Conference Held in Seattle*, Washington, U.S.A., February 1974, DHEW publication (NIH), 75-653; U.S. Dept. of Health, Education, and Welfare, *Public Health Service*, National Institutes of Health, 1975 (publ. 1976); Arthur Kleinman, *Patients and Healers in the Context of Culture: An Exploration of the Borderland between Anthropology, Medicine, and Psychiatry*, Comparative Studies of Health Systems and Medical Care, 3, University of California Press, 1980; Arthur Kleinman, *Social Origins of Distress and Disease: Depression, Neurasthenia, and Pain in Modern China*, Yale University Press, 1986(凯博文著,郭金华译,《苦痛和疾病的社会根源:现代中国的抑郁、神经衰弱和病痛》,上海,上海三联书店,2008)。第一部书包括好几位精神病医生的论文。

> 凡产时当向坐卧,若触犯禁忌,多令运闷,故血下或多或少。是以产处及坐卧,须顺四时方面,避五行禁忌,若有触犯,多招灾祸也。①

这个讨论清楚地表明,这种疾病的终因是坏行为,即违犯了有关四时五行的某些规则。它说这些规则就是禁忌。这种禁忌在古典医书里不常见。这些禁忌表明,这种疾病来自大众医学界。在大众医学里,对于诸如分娩之类的事情来说,面对一定的方向是很重要的。

这并不是胡扯。正如任何现代医生都知道的那样,眩晕、作呕和血流量变化是紧张和焦虑的常见症状。让我们设想很早以前,一个接生婆正在为难产妇接生。接生婆也许终有一天会相信,分娩困难是因为产妇面朝禁忌方向。如果她提醒她将会得一种危险的疾病,那么结果自然就是格外的焦虑以及这种失调的其他症状。

现在让我们从人类学的观点来检视这种治疗失调的疗法。《本草纲目》(1596)中有一个源自更早资料的方子,这个方子就是用分娩的血治疗叫做"产乳血运"的失调:"取酽醋,

---

① [隋]巢元方,《诸病源候论》卷四十三:

> 运闷之状,心烦气欲绝是也。亦有去血过多,亦有下血极少,皆令运。若产去血过多,血虚气极,如此而运闷者,但烦闷而已。若下血过少,而气逆者,则血随气上掩于心,亦令运闷,则烦闷而心满急。二者为异。亦当候其产妇血下多少,则知其产后应运与不运也。然烦闷不止,则毙人。凡产时当向坐卧,若触犯禁忌,多令运闷,故血下或多或少。是以产处及坐卧,须顺四时方面,避五行禁忌,若有触犯,多招灾祸也。

血气:通常都误认为"血"的意思总是表示身体血液。医学作者把体内的气划分为两方面,一方面是阴,他们称之为血,另一方面是阳,也称为气。见 Nathan Sivin, *Traditional Medicine in Contemporary China: A Partial Translation of Revised Outline of Chinese Medicine*, Science, Medicine and Technology in East Asia, 2, Ann Arbor, University of Michigan, Center for Chinese Studies, 1987, 50–60。

和产妇血如枣大,服之。"①

使用人血的做法再次使人联想到,这个方子取自大众医学,但是这并没有回答人类学家紧接着要问的问题:仪式是什么?今天,人们只不过停一会儿,吞下一粒药丸,还是接着做他们正在做的事。但是,大众法师用的是仪式,即表现极具象征意义的行动和谈话模式。我们并不知道,在这种情况下会使用什么象征和行动,去平息患者的焦虑,中止其眩晕。

我们的确在另一个原始资料中找到了一种仪式。7世纪后期,孙思邈《千金翼方》中有一个很不错的集子,汇集了道士们②用来驱除缠人致病的灵魂的种种方法。这个《禁经》包括了治疗"产运"仪式的细节。从这里,我们可以知道诸如这类仪式是如何生效的,即使它们看上去是不科学的:

> 取蒜七瓣。正月一日,正面向东,令妇人念之一遍,夫亦诵一遍,次第丈夫吞蒜一瓣,吞麻子七枚便止。丈夫正面向东行,诵满七遍。不得见秽恶,守持之法,不用见尸丧,见即无验。吾躐天刚游九州,闻汝产难故来求,斩杀不祥众喜投:母子长生相见面,不得久停留,急急如

---

① [明]李时珍,《本草纲目·人部第五十二卷·人血》。李时珍认为这个方子源自宋《太平圣惠方》,但经检未见此方,方见明《普济方》卷三十八载有此方。库伯和席文的讨论,见 William C. Cooper and Nathan Sivin, "Man as a Medicine: Pharmacological and Ritual Aspects of Traditional Therapy Using Drugs Derived from the Human Body", in Shigeru Nakayama and Nathan Sivin (eds.), *Chinese Science: Explorations of an Ancient Tradition*, MIT East Asian Science Series, vol. 2, Cambridge, Mass, 1973, 203—272.

② 一个方法就含有"吾为天师祭酒"的话语,见孙思邈《千金翼方·禁经上·禁鬼客忤气第六》。

律令。①

最后一句是要诵念的咒禁。最后十个字表达的是道士把恶鬼送上路,又是道士表达其在天官系统中的权威性的惯常方式。

从这里我们可以看到象征治疗的要点。在仪式之前,这个妇女是与世隔绝的、被危险灵魂所纠缠的无助的受害者。仪式使她变成一个被道士保护的人,而道士则拥有驱除甚至灭掉恶鬼的权力。②

但是,这个仪式是如何防止分娩之后眩晕的呢?它能改变的是什么?为了回答这个问题,我们不仅要看那个妻子的想法,还得看她与丈夫的关系。她丈夫为什么在这个仪式中所起的作用事实上要大于妻子的作用呢?

重要的是,在 7 世纪,事实上直到近几十年,当一个汉族女人结婚了,通常就搬到她丈夫的父母家去。至少到生了个男孩为止,她主要的责任是侍奉他们。传统社会并不鼓励丈夫去考虑孕妇妻子的焦虑。如果丈夫对其妻子显露出过多的钟爱,他可能就会冒犯其父母。

这个仪式使丈夫的注意力聚焦到他妻子分娩时将要面临的危险上。诵咒禁让他公开地表达他的关怀。如果他真心实

---

① [唐]孙思邈,《千金翼方·禁经上·咒禁产运第十》。关于天柱,即把星星绑在一起的宇宙子午线,见 Edward H. Schafer, *Pacing the Void T'ang Approaches to the Stars*, University of California Press, 1977,第 12 章。

② 对类似情况的经典研究,见 Claude Levi-Strauss, "L'efficacite symbolique", *Revue de l'histoire des religions*, 1949, 135: 5 - 27. 英译文见 "The Effectiveness of Symbols", *Structural Anthropology*, New York, Basic Books, 1963, 186 - 205. 中译文见:克洛德·列维-施特劳斯,《结构人类学》,张祖建译,北京,中国人民大学出版社,2006,第 197 - 219 页。

意地这样做了,那么,结果自然就是他妻子对分娩就不怎么焦虑了。无疑,有些丈夫表明他们是关心的,但也有丈夫并不关心。至少,这样的仪式可以促进他们改善关系。换言之,结果不仅是心理上的变化而且还有社会性的变化。早期中国,难以想到还有鼓励这种同情关怀的其他方法。

结　语

在记载了大众医学的许多文献之中,有些文献,比如像刚才谈到的这些,可以用人类学和社会学的工具进行分析。这对于医学史家来说是创新的工作。除了通常的训练之外,他们需要学会医学人类学和医学社会学的技能。作为从事此项工作的学者,我们将开始理解古代中国绝大多数人的卫生保健,这些人没有机会接近精英医生。

同样的方法对于传统工艺的研究也是有价值的。大众文化是农业、制陶、金属工具制造以及许多其他行业的基础。最近在自然科学史研究所做研究的曹圣洙一直在研究这些领域的工匠文化以及明清时期宗教在其社会关系中的作用。[1]

访谈是另一种成熟的社会科学的工具,也为揭示大众文化的最近的历史提供了很好的机会。大概在中华人民共和国的头 15 年,大众文化以很多传统的方式延续着。很多传统的

---

[1] Philip S. Cho, *Ritual and the Occult in Chinese Medicine and Religious Healing: The Development of Zhuyou Exorcism*, Ph. D. dissertation, History and Sociology of Science, University of Pennsylvania, 2005.

行医业务和工艺实践都一直在按种种方式开展着,这些方式是值得了解的。1965年,活动着的法师还在接受许多仪式训练,这些仪式自那时起就被遗忘了。有一定数量的这类从业者和工匠仍然活着,但是他们很老了。如果没有人记录他们所记得的东西,那就来不及了。

在社会学和人类学里,访谈的方法不仅仅是开始对话和打开录音机。人种志访谈是一门拥有自身有力方法的学科。想做访谈的人可以求学者教,或者从书中学。[①]

在科学史、技术史和医学史里,大众文化是一个重要领域。在中国,有些人对做这样的研究感兴趣。对于一般的史学家来说,意识到大众文化多么重要,并且出于自己的目的对之进行研究,也是明智的。要理解中国史,就需要思考所有中国人的过去,包括都市的和乡村的,识字的和文盲的,汉族的和非汉族的。对于这种广泛的探究来说,人类学和社会学是经受过检验和验证了的工具。

---

① 我发现James P. Spradley 的 *The Ethnographic Interview* (New York: Holt, Rinehart and Winston, 1979)用来自学很不错。近期的手册是Steinar Kvale, Svend Brinkmann, *Interviews: Learning the Craft of Qualitative Research Interviewing*, 2d edition, Los Angeles, Sage Publications, 2009。关于录音的使用,可读Daniel Makagon 和 Mark Neumann 的 *Recording Culture: Audio Documentary and the Ethnographic Experience*, Los Angeles, Sage, 2009。

第四讲

比　较

  史学工作不仅比较两种不同文化中的某个东西,也比较同一种文化中处于不同时期和不同地域的概念或者习俗。"自然"与"nature"、"身体"与"body",其含义并非一一对应,而且,它们各自在中国和欧洲历史的不同时期的含义也大为不同。西方天文学和医学在中国和日本的影响大不相同。理解一个国家对待外国科学的态度,需要考虑政府变化、新旧方法的价值、社会组织以及学习动机之类的因素。中国人对待轮船和蒸汽机车的不同态度,反映出他们对于网络技术对社会影响的认识,以及官员们聪明的外交手腕。

许多科学史学家把比较用作学术工具。人类经验的可能性如此之多,以至于任何一种文明都只利用了这些可能性中的一小部分。一位在法国长大并且只研究法国史的学者,很可能想象不到在亚洲和非洲所实现的那些可能性。这是令人遗憾的,因为这些可能性也许给她提供了理解法国文化的思路,而这种思路可能是她除此之外永远都不会知道的。

阅读中国的期刊就很清楚,这个国家与其他国家一样,比较对于史学家来说通常是有用的。这次报告谈的是值得思考的有关比较的几个问题。

先 决 条 件

许多关于比较的出版物根本没有影响,其原因值得思考。对于正在思考如何在自己的工作中运用比较的一些人,我需要提醒几点。首先,尽管我们经常比较两种不同文化中的某个现象,但这并不是唯一的方式。很多史学工作实际上就是比较同一种文化在不同时间和不同地点中的某个理念或者习俗。我确信这里的每个人都知道,中国女性精英在唐朝的自由,到明清时期已经失去。后两个朝代,多数女性精英与世隔绝,没受过教育,因裹脚而半残。另一个例子,如果比较北宋前后的行医,我们就可以看出,大约在11世纪,行医的方方面面差不多都发生了大变化。①

而且,我们想到比较时通常就会想到两种文化。做这种研究,必须熟悉两种文化。然而,我们经常看到的,却是一种文化方面的狭窄专家在做两种文化的比较探究。这种探究往往由于一个简单的理由而导致不均衡的结论。你所熟知的一种文化总是比你略知一二的文化看上去更加丰富、更加充满迷人的理念和活动。如果你已经研究伊斯兰文化多年,仅仅阅读几本关于印度的书,你未必会认识到关键的相似之处和不同之处是什么。即使你阅读的书很优秀,这些书的作者也未必就提出了你想回答的问题。要想做出均衡的评价,你还是需要同等地精通这两种文化。协作就是一种方法。为了富有成果的协作,两人都需要对彼此的领域有不少的了解。

要熟知任何社会,都需要懂得其语言。任何希望高质量的期刊或出版社接收其著作的人,都得使用原始资料,并且还要知道其所有前辈的工作。我校研究生如果想研究中国古代的数学,他们不仅需要懂文言和白话文,而且还要懂法文和日文,因为很多重要的著作都是用这些语言写就的。同样的理由,医学史学家需要懂德语。

另一方面,如果你对研究伽利略(Galileo Galilei,1564—1642)感兴趣,你就必须熟悉伽利略写的拉丁文和意大利文,

---

① 关于医学和其他领域的变化,见孙小淳和曾雄生编,《宋代国家文化中的科学》,北京:中国科学技术出版社,2007年。其中8篇论文取自2006年在杭州举行的"宋代国家与科学国际学术研讨会"。

它们不同于古典拉丁文和现代意大利文。你还要懂现代意大利文、法文、德文和英文,因为有很多重要的研究使用的是这些语言。具备这些先决条件是很难的。实际上,在进研究生院之前还没有开始学习所需要的语言的人,会发现学这么多语言太晚了。但是,花了好几年做一篇学位论文,把它修订成一本书,却发现它不能出版,是再糟糕不过的了。另一方面,一旦你对此付出努力,你的研究可能就处于世界上最佳者之列了。

比较的例子

我将列举一些比较的例子来说明它可能用得多么广泛。一些例子将会说明广泛研究使得解释你所发现的差异更加可能。我先从用字典做意义比较开始。

并非每个学生都知道如何使用所谓"历史性词典"。对某个词的每一个意义,这样的参考书都引用了最早用法的例子。其编者不能对每个词做原始探究,但是如果学者们批判地使用它们,它们就是极有帮助的。英语的标准字典是《牛津英语词典》,多数学者将它简称为 OED。可以通过很多大学图书馆在线使用它。也有几种这样的汉语词典,包括优秀的《汉语大词典》。

自　然

使用这种书,会让你发现重要的比较问题。"自然"就是一个例子。普通汉英词典通常把它定义为名词"nature"——因为今天人们把它用作与"大自然"相当的词——或者用作形容词"natural"(自然的)。如果在《汉语大词典》里查阅"自然",我们只能查到四种古代的意思[1]:(1)天然,非人为的,"natural"(出自《老子》)[2];(2)不勉强,不拘束,不呆板,"unforced"(出自《后汉书·郎𫖮传》);(3)不经人力干预而自由发展,"spontaneous"(出自叶适《台州商君墓志铭》);(4)犹当然,"naturally"(出自《北史·裴叔业传》)。根据此词典,古代"自然"的意思没有包括作为名词的"自然"含义。

为什么没有包括呢?显然是因为当时中国并不需要一个表示物理世界的词。为什么我们要假定他们需要它呢?事实上,一项关于现代中文词起源的研究揭示,最先用"自然"表示"nature",是在1881年的一部日语哲学术语字典的译本中。[3]

---

[1] 汉语大词典编辑委员会、汉语大词典编纂处编纂,《汉语大词典》(第8册),上海:汉语大词典出版社,1991年,第1328页。

[2] 在含义(1)之下给出的三个例子,都归结为"天然"的意思太模糊。另一方面,说成是"非人为地"才恰当。因为这个词的字面意思是"然",而没有"自"的意思。我对这种用法的理解是它有别于含义(3)。我发现,自唐以降,"自然"才有"自然的"清晰意思。

[3] 迈克尔·莱克勒(Michael Lackner)及其助手做的这项研究尚未发表,查阅 Michael Lackner and Natascha Vittinghoff (eds.), *Mapping Meanings: the Field of New Learning in Late Qing China*, Sinica Leidensia, v. 64, Leiden: Brill, 2004。亦见 G. E. R. Lloyd and Nathan Sivin, *The Way and the Word: Science and Medicine in Early China and Greece*, New Haven: Yale University Press, 2002, pp. 199-200 中的讨论。

总之，在一本好词典里查阅"自然"，揭示了它直到一个世纪多一点之前才有"nature"的含义。如果我们在《牛津英文词典》里查阅"nature"这个英文词，我们就会知道，直到1400年它才有大自然的含义（第11个意思）。① 中文和英文概念都有历史。两个词相比较的历史或许会形成一个有意思的研究项目。

身　体

在最后一个例子里，我们看看一个中文词。现在让我们来看一个英文词"body"。其主要定义是指人、动物或植物的整体结构。中文有好几个词常常被译为"body"，但是它们当中没有一个是表示人、动物或植物的整体结构的意思，而且它们在意思上的彼此差别很大。

让我先从"身"这个表示身体的日常中文字开始。孔子的一位门徒曾子说，他"日三省吾身"，不是用镜子照自己。实际上他是在自问"为人谋而不忠乎，与朋友交而不信乎，传不习乎"②。他的"身"是指自己的言、行、思。今天，"身份"是表示"identity"的一个中国官僚术语。1500年以来，"身份"指的又是社会出身和地位。它与高矮强弱也没关系。换言之，"身"的含义比英语单词"body"要宽泛，它包括许多与肉体无关的

---

① J. A. Simpson and E. S. C. Weiner eds., *The Oxford English dictionary*, vol. 10, 2nd ed., Oxford: Clarendon Press, 1989, p. 248.

② 《论语·学而》。

理念。

欧洲学者译为"body"的其他文言词,都不同于"body"意思的范围。"体"、"形"和"躯"就是其中的三个。在本文附录里,我给出了其含义的例子。大家可以从中看到,体通常是道德品质的某种体现。"形"通常指的是非肉体的特征。"躯"通常的含义不是身体而是生命。

甚至在医学作品中,这四个词的含义都没有以"body"表达的方式把肉体与各种非肉体的东西——心智的、精神的和社会的——分离开来。一位两千年前的中国人,只要他愿意清楚,他就可以让人知道,他谈论皮肤伤口时指的不是道德理念。但是,他的"身"不像西方医生所看见的"body",其中,皮肤、活力、情感和伦理是同一个事物中同等重要的诸部分。这些中文词所谈论的与西方"body"所谈论的就不是同一回事儿。这就是我曾在新加坡做一次题为"中国人缘何无身体"的演讲的原因。

为什么中国人的"身体"与欧洲人的有如此大的差异?部分答案源自公元前250年以后形成的综合哲学。正如我在第三讲里解释的,一个良治的国家和一个健康的身体,都与宇宙的规则循环相和谐。宇宙不仅是物质的,又是一种机体和一种道德秩序。国家亦如此,人体亦如此。

作为这种关系的一部分,直到现代,"身"的重要方面,仍然是心目中的典型以及与他人的关系。这些特征易于被人们忽视,但是比较使它们引人注目。它们是一个非常特殊的历史进程的产物,仅仅发生在中国。

东亚使用的天文学和医学

有些欧洲的史学家曾经论证说,中国人和日本人接受西方科学缓慢是由于对外国人的轻视仇视。这并非见识广博的判断。如果我们分别来看科学的两个方面,比如说天文学和医学,显然中国人和日本人以非常不同的方式对它们做出了响应。要理解这种差异,我们必须既审视技术工作又审视社会组织。

稍在1600年之前,来自欧洲的耶稣会传教士抵达日本和中国。其目的是让这两个国家的统治阶级都皈依基督教。他们很快就发现,这两国统治阶级中的多数人并不愿意受外国宗教的诱惑。为了吸引他们的注意并使他们相信西方人是儒者,他们就撰写天文学、医学及其他技术专题的书籍。结果,17世纪的中国就对外国的天文学作品极感兴趣;但是医学作品毫无影响。另一方面,日本人显示出对西方医学作品的极大兴趣,但是对天文则不感兴趣。我将对做出如此不同响应的这两个社会的境遇进行比较。

中　　国

中国晚明时期,诸如著名的徐光启(1562—1633)之类的高官,把传教士们的数学和天文学作品引进皇宫。帝王拒绝委派外国专家到司空监任职,不许出版他们的作品。清初之前,中国人中只有很少的非官人员研究外国的天文学,并且评价其优点和弱点。他们的作品极有趣,但对其他人几乎没有产生什么影响。

满人入侵中国的时候,传教士帮助他们铸大炮。他们使满人相信,传教士能帮他们统治中国人。两周之内,满人的军队长驱直入北京,新的统治者任命了一个欧洲人为钦天监监正来表达感激。新的监正抛弃做天文工作的多数传统方法,采用了欧洲人的方法。我们可以得出结论说,耶稣会士的天文学对中国产生了影响,因为清政府正式采纳了欧洲人的方法。①

对耶稣会士的医学作品则没有这样的响应。至少有一本解剖学书籍《泰西人身说概》出版于满人入侵之前。但是印数不多,也没有广泛传播。另一个原因是,1640年前后的欧洲医学并不优越于中国。解剖学是其最新奇的部分,但当时它在西医中还没有实际应用,中医也没有实际用过它。②

日　　本

在日本的境遇就完全不同。就医学而论,我们必须理解这个国家的社会结构。日本统治者试图维持一切工作全都靠世袭的一种社会,没有人能改变职业。如果你父亲是位厨子,人们会期望你也当厨子并且传承你父亲的厨艺。这是复杂社会的非同寻常的结构。这也是一个困难的生存结构。你不能

---

① 西方人在中国的影响,以及中国人在欧洲的影响,见韩琦的多种作品。例如,韩琦,《中国科学技术的西传及其影响(1582-1793)》,石家庄:河北人民出版社,1999年;韩琦,关于十七、十八世纪欧洲人对中国科学落后原因的论述,《自然科学史研究》,1992年,第11卷第4期,第289—298页;韩琦,17、18世纪欧洲和中国的科学关系:以英国皇家学会和在华耶稣会士的交流为例,《自然辩证法通讯》,1997年,第19卷第3期,第47-56页。

② 马伯英、高晞、洪中立,《中外医学文化交流史:中外医学跨文化传通》,上海:文汇出版社,1993年,第281-286页。

合法地中止厨子工作,不能合法地从事另一种职业。

这种刚性体制使人安分守己时,还没有医生职业。各种人都在疗病,其中很多是佛僧。中医在7世纪就进入了日本,但是很长一段时间主要在宫廷和官府使用。大约1400年以后它才缓慢地在社会上传播。到1700年,才有足够多的医生给普通人做治疗。不久,荷兰商人开始传播少量西医知识。1774年,一个日本人团体出版了一本译自荷兰文的解剖学插图书。①

这本书极有影响。它在治疗上虽然没有实际用途,但在日本社会非常重要。外国医学知识进入日本一个多世纪的时候,人们发现它们创造了一个新的职业。当医生虽然是不合法的,但是新医学履行着一项有价值的服务,并且是政府允许的。一个农夫可能会从乡村进城,师从某位老师学医,并自称是医生。到他退休时,他可能很富裕并且有许多弟子。实际上,1800年之前,他可能对西方疗法知之甚少,但是他在学解剖学的时候其资质与其他初学者是一样的,他的学识赋予他以权威。他会开始运用他所知道的那些技能。欧洲医学在日本的流行,是因为它向那些别无选择的人们开创了一种职业。

但是天文学在日本却是另外一回事了。皇宫使用中国天文学,有一位宫廷天文学家。日本贵族对精确历法的兴趣不及占星术和占卜,因此世袭的阴阳师并不擅长天文学。甚至在17世纪的将军幕府,至少雇佣了一位一流的数学家来做天文工作,没有理由去发明新方法。中国取代授时历很长时间

---

① Sugita Genpaku, *Dawn of Western Science in Japan: Rangaku Kotohajime*, tr. Ryôzô Matsumoto, Tokyo: The Hokuseido Press, 1815/1969.

之后,日本政府还在继续使用它。换言之,在日本没有对外国天文学好奇的动机。

这个例子使我们想到,如果我们想理解对待外国科学的态度,仅看态度或文本是没用的。对中国和日本的比较提醒我们,诸如政府的变化、旧新方法的价值、社会组织以及学习动机之类的事情,都至关重要。要做出有价值的比较,明智的做法是知晓社会、政治、行政、智识、经济等历史维度。下一讲我将回到这一点上来。

网 络 技 术

我曾经问过,中国人为什么不像欧洲人那样,在19世纪居然拒斥蒸汽动力。我曾与过去的一位来自武汉的学生张忠一起做研究,他早前做博士论文是研究这些问题的。[①] 这与前面的那个例子中的问题不同。我们的结论是,必须考虑恰当的技术问题。

有些西方史学家论证说,反动的中国制度、意识形态和思考习惯使它不可能在20世纪之前接受现代技术。这个论点假定,如果传统没有中止它们,理性的人们就会渴望变革(他们不必是非理性的)。在我看来,这并非一个明智的解释。如果我们考虑19世纪后期的国际政治是多么复杂,一个好想法

---

① Nathan Sivin & Z. John Zhang, Railways in China, 1860−1898: The Historical Issues, *History of Technology*, 2004, 25: 203−210; Zhang Zhong, The Transfer of Network Technologies to China: 1860−1898, Ph. D. dissertation, History and Sociology of Science, University of Pennyslvania, 1989.

就是谨慎地探究什么是合理的、什么是不合理的。

政治环境

1865年,由于太平天国运动等反抗运动和鸦片战争,清政府失去了很多大权和收入。1840年到1949年,中国甚至连一年的和平和安全都没有经历过。19世纪后半叶,鸦片战争失败之后,中国的法律不再治理其领土各部分,外国势力可以通过威胁发动另一场战争来强迫中国政府做任何事情。另一方面,他们不能随意制造这样的威胁。他们不准备征服和统治中国;他们想要的是可靠的商业收入。当贸易受到威胁时,就意味着要使用外交和战争手段,但是得用不再威胁贸易的方式使用这些手段。因此,有外交手腕的聪明华人官僚就可能阻止欧洲势力的联合。

中国人对外国科学技术——或者其他东西——并非都意见一致。高官们甚至反对几乎所有的出口。另一方面,许多皇族成员和汉族高官都反对接受西方的一切。他们认识到西方的价值观、制度和机器会破坏中国文化和他们自己的权力。

但是另一些人认识到,欧洲技术的背后有重要的理念。诸如握有大权的官员曾国藩(1811—1872)、李鸿章(1823—1901)、左宗棠(1812—1885)和张之洞(1837—1909),从鸦片战争知道了采用外国技术是抵御外国人的唯一可能方式。大约1860年以降,他们购买外国的机器、雇佣欧洲人造机器并训练使用者。1874年,李鸿章提出,必须有铁路和电报来抵御海军袭击,朝廷允许他们为军事目的修筑铁路。

蒸 汽 动 力

我们问的是蒸汽动力问题。我们发现,中国官员对不同种类的蒸汽机的反应方式完全不同。1875年之前,他们热切地购买并开始建造轮船。另一方面,他们坚定地抵抗着让欧洲人修筑铁路的巨大压力。政府对待蒸汽机车的政策与对待轮船的政策根本不同,而与对待电报线路的政策相同。这看似奇怪,但是张忠找到了一种解释。

网 络 技 术

技术史学家托马斯·休斯(Thomas Hughes)在其《力量网络》[1]一书中说明,尽管铁路为了远距离运输使用了蒸汽动力,但其发展进程与轮船却没有共同之处。另一方面,它与电报以及以后出现的电力网络倒是非常相似。

铁路是网络技术。它以持续和可视的方式延伸了钢轨。它占据土地,因此捣乱者易于阻止。如果某人破坏了铁轨,火车不可能行驶了。它需要标准的轨道尺寸。当线路变长,其成本就降低。修筑者总是可以通过扩展网络来创造更多的利润。

在中国的欧洲商人对这些情况很了解。认识到网络技术的这些特征并基于这种理解来做决策,并没有花李鸿章等人多少时间。让我们看一个例子。

外国商人知道,最重要的一步是修筑一段几百米长的铁

---

[1] Thomas Parke Hughes, *Networks of Power: Electrification in Western Society, 1880–1930*, Baltimore: Johns Hopkins University Press, 1983.

路线。他们有把握，铁路一旦存在，中国人就不会阻止他们把路线加长。与外国人打交道的中国官员很快就意识到，网络一旦开始，他们不能阻止扩建了。在多数中国官员看来，外国人拥有铁路和电报是一种潜在的侵略手段。这个看法完全正确。

1876年，一个美国外交官组织了一个公司来修筑一条马路。完工时，它成了一条一公里长的铁路，还带有一个小机车。数月之后，所有者扩建了铁路达六公里长，并且给了一个较大的火车头。由于在没有另一场危险的战争的情况下，政府既没有毁掉它也没有阻止它的延长，因此就用很高的价格买下了它并把它搬到了台湾岛。

因为当时台湾不需要铁路，史学家们通常就把它描述成不合理的决定。但是当时在台湾，一个网络未必会受外国人掌控并延伸。无论如何，它对于入侵来说是没有价值的。商人们认识到，他们不能以计击败他们的中国对手，1895年以后就停止了修筑新网络。是年，日本在军事上战胜了中国最后的防卫。

在这个例子中，理解了蒸汽动力不是一种单一的技术，就可能明白19世纪晚期中国官员的推理，就能意识到他们在那个时期的环境下尽力而为的能力。总之，比较中国和西方对蒸汽动力的使用，导致了对中国人外交技能的更好理解。

结　语

我已经给出了几个如何比较的例子,而周全地思考和批判地比较,在解决历史难题中是有用的。关键是理解所有的情况。这使我们有可能弄清楚为什么有些方面是相似的,而其他方面则是不同的。这也能解释比较的结果是真正有用的。

我的下一讲将论述文化簇,这是另一种较可靠的评价工具。

附　录

表达"Body"等事物的一些中文词

词	主要意思	宽泛使用的例子
身	身体,自身,人身,身份	吾日三省吾身:为人谋而不忠乎?与朋友交而不信乎?传不习乎?(《论语·学而》)医学用法,见下面的"形"
体	身体,某些道德品质的具体化,体现	阴阳合德,而刚柔有体(《易经·系辞》)
形	形式,形状,特性,形体,体型	金形之人……其为人……身清廉……(《皇帝内经·灵枢》)
躯	身躯,躯体,生命之躯	其为人也小有才,未闻君子之大道也,则足以杀其躯而已矣(《孟子·尽心》)

Chapter 5

第五讲

运用"文化簇"概念

　　文化簇进路就是用所有相关学科来考察所有的相关资料,理解人文或社会科学问题。这种进路从给定问题存在于其中的文化的所有维度,如科学、观念、社会关系、经济、宗教、政治、亲属关系等等,来研究该问题。用这个想法可以发现,《授时历》受外来影响很小,是因为忽必烈保持不同族群之间的知识分离,好让自己享有多种选择的特权;北宋博学之士中,有科学思想者都担负着与科学相关的公务职责,而明末晚清的博学之士在科学上有创新精神者,多是考证运动中的名师;对"李约瑟问题"数以百计的不同回答没有导致任何有用的结论,其原因在于答者不明白科学革命并非一定导致社会革命;不二臣的惯例在宋—元转型期不重要,但在明—清转型期却盛行起来,这种反差是一个有待研究的有趣问题。

第五讲 运用"文化簇"概念

我今天要谈的是"文化簇"(cultural manifolds)①。你们中的一些人可能已经发现,它是历史研究的一条有用进路。中国的一些学者,包括北京的孙小淳和上海的董煜宇,已经在使用这种方法②。其含义实际上很简单。

多数学者是作为社会史、思想史或其他学科的专家受到训练的。他们把自己限制在一个历史维度上,试图达到尽可能的深度。我们所读到的多数学术作品就是由这种专注所产生的。

除了给其他领域的专家撰写的范围狭窄的论著之外,也需要涵盖较宽领域的研究并与更多专业领域的人士进行交流的著作。尤其是,仅仅着眼于科学技能或者社会关系,很多问题是回答不了的。有时二者都必须考察;有时则必须从不同

---

① "Cultural manifold"是罗维(此前一直被学界译为"劳埃德")和席文合著的一本书(G. Lloyd & N. Sivin, *The Way and the Word: Science and Medicine in Early China and Greece*, New Haven: Yale University Press, 2002)中提出来的一个概念。虽然此前也有人使用这个概念(见任定成等,"文化簇"与中国科学技术史研究,《中国科技史杂志》,2010年,第31卷第1期,第14—25页),但将此概念用于科学史,罗维和席文是第一次。席文最初将之译为"文化整体"(见席文,文化整体:古代科学研究之新路,《中国科技史杂志》,第26卷第2期,第99—106页)。孙小淳译之为"文化多样体"(见孙小淳,走近希腊科学:读劳埃德的《早期希腊科学》,《科学文化评论》,2005年,第2卷第3期,第115—125页)。任定成将其译为"文化簇"(任定成等,前引文)。——译者

② 见孙小淳和曾雄生编,《宋代国家文化中的科学》,北京:中国科学技术出版社,2007年,尤其是两位编者在其中的论文;又见董煜宇,《北宋天文管理研究》,博士论文,物理学史,上海交通大学,2004年。

的学科,比如经济学或宗教学来考察。文化簇的理念,就是用所有相关学科来考察所有的相关资料。

学科不是历史实际的界限,而是大学的系科。如果你是一位物理史家,人们很可能这样教过你,即物理科学是专家研究的东西,其他任何东西都只是史境(context)[①]。但是在历史实际里,任何东西都同样真实,没有什么东西仅仅是史境。

我在这次演讲中把所描述的方法称为"文化簇",因为在数学中,一个簇存在于多个维度上。我的话题不是数学而是历史。我们感兴趣的是与给定问题及该问题存在于其中的文化有关的所有维度(理念、社会关系、经济、宗教、政治、亲属关系等等)。把这种方法的名称译成汉语并不容易。把它描述成"历史多法"或许更简单些。这样就清晰了,但是不能解释学者为什么使用它。这种描述还会给出一个错误的想法,就是以为它仅适用于做历史研究。实际上,它是一条可以用于任何人文或社会科学学术研究的广泛的进路。有些人也许认为它就是一种文化史。这或许有理,但我不这么看。"文化史"太含糊,难以把握。文化簇的使用在一般的历史研究中并不新,但是用它对中国科学史和医学史进行研究的工作却不多。我将给出一些用于这种目的的例子。

---

① 中国语言学家将"context"译为"语境",范岱年译之为"与境",任定成将其译为"史境",台湾学者将其译为"脉络"。——译者

伊斯兰对中国天文学的影响

首先,在我最近的一本论述授时历的书中提出了一个有趣的问题。授时历是中国历史上最先进的历法(数理天文学)体系①。这个体系完成于1280年,是为了庆祝蒙古人的胜利,蒙古人通过长期厉害的战争征服了中国南方。自忽必烈成为年轻皇帝以前,就已经有一群汉人顾问,他从他们那里懂得了汉文化。他通常挑选擅长阴阳术的人作顾问,他们的阴阳术指的就是天文学、占星术和占卜。

他想劝说中国南方政府投降。其目标是用一种和平方式统治重新统一了的中国,使征税变得容易。他的顾问们劝他说,有必要设计出汉人能够理解并且接受的政府模型。刘秉忠(1216—1274)等顾问在北京建了都,并且让城充斥着汉式官衙。他们在古代王朝的基础上设计了国家礼仪体系。根据忽必烈关于汉人臣民的观点,这个政府是一个传统的政府。

实际上,在蒙古人的统治下中国被划分成了四个法定等级。蒙古人本身就是统治者,处于顶层。贵族非常有权势。他们可以要求政府让汉人做奴隶,要求拥有带农户的农场。由于蒙古人普遍不能读写,所以他们的治理就依赖第二等级

---

① Nathan Sivin, *Granting the Seasons: The Chinese Astronomical Reform of 1280, With a Study of its Many Dimensions and a Translation of its Records*(《授时历丛考》), Sources and Studies in the History of Mathematics and Physical Sciences. Secaucus, NJ: Springer, 2008. 请参阅其中的详注和参考文献。亦见席文,为什么《授时历》受外来的影响很小?,黎耕译,《中国科技史杂志》,第30卷(2009年)第1期,第27-30页。

的非汉人,他们称之为色目人①。色目人是受过教育的维吾尔人、西域人、西亚人、中东人以及少数欧洲人。他们做记录、征税,以及其他治理工作。在中国北方,处于第三等级的是汉人和其他早期被蒙古人征服的人。南方人1276年才成为蒙古人的臣民,处于低等级的最底层。汉式官僚机构的权力高不过头两个阶层。它只能统治汉人,主要是中书省②,即京城周围(腹里)的那些人。只有皇帝认可了汉官们的建议时,汉官们才能排除蒙古人的相反意见。换言之,尽管从下看,朝廷似乎就是汉人,但这是假象。从上往下看,朝廷则是蒙古人。

我心里想探究的正是这个问题。元朝的中国是一个非常庞大的国际性帝国的一部分。忽必烈统治的仅是一部分。他的政府雇佣来自伊朗和中亚的穆斯林天文学家,甚至在北京还有回回司天监。1275年,政府把回回司天监和汉儿③司天监合并起来。

人们期待的自然是,当1276年开始研究新天文体系时,该体系就会很好地得到关于穆斯林天文学的知识。因此,史学家们就留意穆斯林对郭守敬(1231—1316)及其同事的工作的影响。结果他们发现这种影响很小,一点儿也不重要。问题是:为什么不重要?

中国、日本和西方的多数中国天文学史学家都期待发现

---

① 色目人,字面意思是"各色名目之人",元代中国西部民族的统称,包括蒙古人、汉人和南人之外的一切中国人,主要是维吾尔人和吐蕃人。其地位在蒙古人之下、汉人和南人之上。——译者

② 中书省是元代中央的最高行政机关。中书省直辖地区称为"腹里",包括黄河以北、太行山东西,即现在的河北、山西、河南、山东及内蒙的一部分。——译者

③ 汉儿,中国古代少数民族对汉人的称呼。——译者

重要的影响,非常努力地寻找这些影响。他们做出了很多研究计划①,今天没时间讨论它们了。最缜密的计划是薄树人制定的②。他认为这种影响微乎其微,而且并不重要。

如果我们问为什么真是这样,那就会有好几个可能的答案。答案之一就是,中国人没有认真对待外国学问。很容易证明这个想法是错的。另一种答案是,尽管回回司天监和汉儿司天监合并了,但任何一个司天监在改革中都没有起重要作用。这个答案证明是相当正确的。忽必烈把改革移交给了一个新的机构,这个机构称为太史院。领军天文学家并非来自旧机构。事实上,他们多是忽必烈的擅长天文学、占星术和占卜的顾问。汉儿司天监变成了为太史院培养低等工作者的学校。据我们所知,司天监里的穆斯林官员和汉族官员在改革中没起直接作用。很常见的是,在中华帝国的天文学机构中——事实上在所有行政机构中——地位低的官员不仅不干什么活儿,还妨碍有能力和努力工作的人。③ 忽必烈创建这个新机构就是保护他的专家们免受这种干扰。

但是,如果京城里有穆斯林天文学家,为什么涉身改革的那些天文学家不向他们学习呢?如果这个问题不属于心理方面的,那是哪个方面的呢?在思考了这个问题的大量答案之后,我最终认识到我不能回答这个问题,因为我原认为这个问

---

① 我已经对这些计划中最有趣的部分做了分析。见 Sivin 2008,前引书,pp. 218-225。

② 薄树人,试探有关郭守敬仪器的几个悬案,《自然科学史研究》,1982 年第 1 卷第 4 期,第 320-326 页。又可见陈美东,《中国科学技术史·天文学卷》,北京:科学出版社,第 528-529 页,2003 年。

③ 著名的例子见董煜宇 2004,前引书。

题是天文学上或科学上的。一旦我从政治维度去探究,回答这个问题就不难了。

缺少影响的原因原来是蒙古统治者治理政策的事儿。如果阅读元朝早期的御膳菜谱《饮膳正要》这部中国最著名的烹饪书,就会发现有很多煮狼头之类菜肴的烹饪法。你很快就能认出多数烹饪法不是做烹饪汉人食物的。这本书包括了很多被蒙古人征服了的那些国家的菜谱①。胜利者喜欢吸收所有这些烹饪方法,而不是把它们变成一种统一的烹饪风格。

同样,忽必烈极热衷于占星术和占卜。他喜欢用各种不同体系的占卜法,以便他能够在它们之间选择最满意的。他想要的是多样的占卜,就像他和其他蒙古人想要多样的食物一样。

就像很多人所知道的,忽必烈的顾问们敦促他恢复元初停止了的科举考试。他拒绝这样做,是因为他喜欢选择自己的官员,或者一个一个地批准他人的选择。

这就是为什么忽必烈阻止许多不同族群的臣民彼此交换信息的原因。当然,他不想他们联合起来反抗他的统治。但重要的是,他想亲自从多种可能性,而不仅仅是从一种天文体系提出的可能性中做出选择。他更愿意通过阻止伊斯兰天文学家与汉人天文学家之间的合作和交流来保持知识的分离。

明朝初年,明太祖就知道京城有大量阿拉伯和波斯天文学书籍。他立即下令学者们考查和鉴定。然后命令他们译成汉文,使汉人天文学家阅读和学习。事实上,我们还有其中的

---

① Paul D. Buell and Eugene N. Anderson, *A Soup for the Qan: Chinese Dietary Medicine of the Mongol Era as Seen in Hu Szu-Hui's Yin-Shan Cheng-Yao: Introduction, Translation, Commentary, and Chinese Text*, The Sir Henry Wellcome Asian Series, London: Kegan Paul International, 2000.

两三部明初译著①。穆斯林和中国的天文学家彼此交流所需要的是一个新朝代。

博学之士

我在前面几次演讲中已经提到了科学史中最令人感兴趣的一个人物,沈括(1031—1095)。他的兴趣异常广泛②。他拥有的原创思想所涉及的主题从地质学到天文学到绘图法到语言学,再到礼仪、音乐、外交、军事、医学、绘画、诗词、茶以及许多其他事物。我总结到,我们只有认识到他的政治生活、行政事务、个人经历和技能是分不开的,我们才能解释他范围极广的兴趣。

他并不是独一无二的。北宋的其他学者型官员的兴趣也非常广泛。任何一位研究科学史的人都知道燕肃(991—1040)和苏颂(1020—1101)这两个人。而这两个人只是北宋

---

① Kiyosi Yabuuti(薮内清), Islamic Astronomy in China during the Yuan and Ming dynasties, rev. & tr. Benno van Dalen, *Historia Scientiarum*, 1997, Vol. 7, No. 1, 11 - 43. Bruno van Dalen and Michio Yano(矢野道雄), Islamic Astronomy in China: Two New Sources for the *Huihui li* (Islamic calendar), in *Highlights of Astronomy*, ed. J. Andersen, CXVIII, Dordrecht: Kluwer Academic Publishers, 1998, 697 - 700. Benno van Dalen, Islamic Astronomical Tables in China: The Sources for the *Huihui li*, in *History of Oriental Astronomy. Proceedings of the Joint Discussion 17 at the 23rd General Assembly of the International Astronomical Union, Organized by the Commission 41 (History of Astronomy), Held in Kyoto, August 25 - 26*, 1997. Astrophysics and Space Science Library, 274, ed. S. M. Razaullah Ansari, Dordrecht: Kluwer Academic Publishers. 2002, 19 - 31.

② Sivin, Shen Kua (1031 - 1095), in *Science in Ancient China: Researches and Reflections*, chapter 3, Aldershot, Hants: Variorum, 1995. 此传记结论部分的汉译文,见 Sivin,沈括,段耀勇、郝建设译,广西民族学院学报(自然科学版),第 11 卷(2005 年)第 3 期,第 39 - 44、55 页。

时期与沈括一样有着非常广泛的好奇心的创新型学者中最著名的而已①。从那时起,学者们倾向于在很多领域学习,从古代经典到绘画。北宋时期,他们的兴趣往往包括科学和技术。但是在南宋、元朝和明朝早期,所谓的博学之士就很少将科学与技术包括在他们的作品之中了。

在蒙古人统治下的中国北方,正如我们所看到的,有另外一种重要变化。很多年轻学者研究天文学、占星术、数学和占卜。元朝早期没有科举考试,这样的研究就提供了一种最好的谋生机会。当蒙古人1315年恢复了常规的科举考试时,情况就再次发生了变化。尽管很多中国人以占卜谋生,但是论述科学的新书却不比以前多。

在明朝和清朝初期,科举考试中经常出现关于观测和计算天文学(天文历法)及其他科技问题。这就迫使每个年轻的精英去研究这些领域。引人注目的是,这并没有在莘莘学子之中引发科学活动和科学著述的复兴。②

从明末到晚清,有另外一个重要的变化。一些学者在"考证"运动中着手研究数学、天文、医学等领域。始于顾炎武(1613—1682)的领导,这样的研究繁荣到清朝中期。顾炎武、清初杰出的天文学家梅文鼎(1633—1721)以及之后的戴震

---

① 张荫麟,燕肃著作事迹考,载《张荫麟文集》,台北:中华丛书委员会,1956年,第116-124页。庄添全、洪辉星、娄曾泉主编,《苏颂研究文集——纪念苏颂首创水运仪象台九百周年》,厦门:鹭江出版社,1993年。

② Benjamin A. Elman, *On Their Own Terms: Science in China, 1550 - 1900*, Cambridge, MA: Harvard University Press, 2005; Elman, *A Cultural History of Modern Science in China*, New Histories of Science, Technology, and Medicine, Cambridge, MA: Harvard University Press, 2006. (本杰明·艾尔曼,《中国近代科学的文化史》,王红霞、姚建根、朱莉丽、王新磊译,上海:上海古籍出版社,2009年。)

(1724—1777)等人都论证说,要理解古代经典的本义,就必须使用朴学和科学的工具①。这场运动在许多方面,尤其是在中西数学技巧的结合上,是相当有创新精神的。

总而言之,如果探究博学之士何时何因把科学工作包括在他们的作品之中,我们就发现,这在北宋时期很常见,而且明末以降也是如此。尽管在其他时期有很多年轻人因为希望升官而去学习科学,但这种学习却没有使他们从事真正的科学活动。为什么没有呢?要记住,研究科学往往意味着记忆科学经典。这是收集知识的很好方式,但却不会自然而然地使大量学者实施科学研究。

另一方面,北宋时期,有科学思想的学者的共同点,是他们都担负着与科学相关的公务职责。沈括主要是财政官员,可是也在司天监服务过。燕肃为皇宫制作仪器,苏颂(曾入阁拜相)曾涉身天文改革。由于北宋政府比早前的任何政府都更深地涉及公共卫生问题,官员们研究医学就是自然而然的了。

考证学者一般来自江南。他们中的许多人在书院任教,这些书院在很多方面像现代大学。他们有强烈的使命感。顾炎武等人论证说,明朝垮台被外国入侵的原因就是失去了孔子和其他圣人的真正教诲。他们认为,原因就是学者们用佛教和道教的异端邪说污染了圣人的教诲。因此,有必要恢复原本的教诲,拒斥数百年来被污染了的学问,回到原始文本中去。他们的工具是朴学和科学学问。因此,就考证运动而言,

---

① Elman, *From Philosophy to Philology: Intellectual and Social Aspects of Change in Late Imperial China*, Harvard East Asian Monographs, 110, Cambridge, MA: Council on East Asian Studies, Harvard University, 1984. (艾尔曼,《从理学到朴学:中华帝国晚期思想与社会变化面面观》,赵刚译,南京:江苏人民出版社,1995年。)

贡献于这个目标的许多学者多不是官员,而是在科学史中有重要作用的名师。

总而言之,从南宋以降,科学作为博学之士的一项活动,其作用发生了多次变化。如果考察这个问题的文化簇,我们就会明白,不同的维度在解释这些变化的原因方面起了重要作用。首先,科技问题总是重要的;这并不是一个内外史的问题。另外,政治因素不时发生相当大的变化;谋生模式也发生变化;如果我们对经济学方面知之更多,我们无疑会发现它起了重要作用。充分理解这个问题,有赖于考虑智识认同、个人认同和社会认同的许多维度。

现在让我给出两个比较问题的例子。

"李约瑟问题"

科学史学家们全都批判地考虑过所谓"科学革命问题"或"李约瑟问题"。讨论这个问题的多数人都忽视了最重要的第一步:定义"科学革命"这个术语。大家都知道,对这个术语的现代历史理解来自托马斯·库恩(Thomas Kuhn)的《科学革命的结构》[①]。一场科学革命就是一种特殊的变化,它涉及新

---

① Thomas S. Kuhn, *The Structure of Scientific Revolutions*, 2d ed., Chicago: University of Chicago Press, 1970. (T. S. 库恩,《科学革命的结构》,李宝恒、纪树立译,上海:上海科学技术出版社,1980年;孔恩,《科学革命的结构》,程树德、傅大为、王道还、钱永祥译,台北:远流出版事业股份有限公司,1994年;托马斯·库恩,《科学革命的结构》,费超译,北京:京华出版社,2000年;托马斯·库恩,《科学革命的结构》,金吾伦、胡新和译,北京:北京大学出版社,2003年。) A. David Hollinger, T. S. Kuhn's Theory of Science and Its Implications for History, *American Historical Review*, 1973, Vol. 78, No. 2: 370-393.

方法，涉及重新理解什么是科学问题、如何解决科学问题，甚至涉及重新理解什么算是问题的解决。发生这些变化，很可能是由于出现了一个全新的理论，或者来自这种文化圈外面的某个理论。例如，哥白尼革命的第一阶段发生在哥白尼去世七十多年后的意大利，是由于伽利略(Galileo)著作的出现。哥白尼(Copernicus)在波兰想利用其新宇宙观改进托勒密的古典天文学，但伽利略却用它们抨击亚里士多德的经院哲学——亚里士多德经院哲学是欧洲大学教育的基础。科学革命在法国并非重新发明，而是从意大利输入的。它输入中国与输入欧洲各国的情况不同。事实上，科学革命大约同时进入中国与法国，比进入欧洲其他国家都要早。

一些史学家否认中国科学发生过任何革命。但是如果我们看一看最优秀的中国天文学家，比如王锡阐(1628—1682)和薛凤祚(1620—1680)的工作，我们就发现，他们迅速地批判性地回应了他们从耶稣会传教士的作品中读到的关于数学和天文学技术的信息。许多史学家并没有认识到这一点，因为他们假定科学革命所导致的社会后果必须与它们在欧洲的社会后果相同。显然，新的科学思想在中国并没有导致广泛的社会后果。因此许多史学家认为，中国科学家不能对革命性的科学思想做出回应。这是由于他们没有阅读王锡阐、薛凤祚等创新型天文学家的作品所致。如果阅读了他们的作品，明显的结论就是，革命性的科学变化可以在不造成社会变化的情况下发生。如果中国的真实情况如此，那么别处的情况也是如此。

换言之，社会革命在科学革命以后最终席卷欧洲是反常的。事实上,许多专家已经说明,一些欧洲人用新的科学思想鼓励社会转型,而另一些人则用科学思想阻碍社会转型。①

这个历史问题并不在于中国:为什么社会革命发生在欧洲?但这并不是专门研究中国科学的史学家能够回答的问题。

我鼓励我的同仁系统地考察中国人开始思考西方科学之后所发生的事情。一些同仁尤其是中国的学者,已经这样做了。② 其他人仍然很迷惑,因为中国并没有社会革命,因此他们认为科学革命也是没有的。他们通常假定,中国科学家不够聪明或者过于刻板,直到20世纪才富有成效地利用西方科学。因此他们继续猜想"李约瑟问题"的答案。结果就是数以百计互不相同、互不兼容的答案。它们并没有导致任何有用的结论。③ 我希望学者们不再把时间浪费在无用的假定上。他们的努力最好放到比较17世纪中国天文学家和数学家与欧洲各国科学家的回应上。

"李约瑟问题"的声望是由于狭隘的技术观点。这就是把没有科学革命与没有社会革命长期混为一谈的原因。一旦认识到既考虑科学问题又考虑社会问题是多么重要,这种混淆

---

① Margaret C. Jacob, *The Newtonians and the English Revolution* 1689-1720, Ithaca: Cornell University Press, 1976. Jacob, *The Radical Enlightenment. Pantheists, Freemasons and Republicans*, London: George Allen & Unwin, 1981. Steven Shapin, *The Scientific Revolution*, Chicago: University of Chicago Press, 1996. (史蒂文·夏平,《科学革命:批判性的综合》,徐国强、袁江洋、孙小淳译,上海:上海科技教育出版社,2004年。)

② 韩琦,《中国科学技术的西传及其影响》(东学西渐丛书),石家庄:河北人民出版社,1999年。黄一农,汤若望与清初西历之正统化,《新编中国科技史演讲稿选集》,台北:银禾文化事业公司,1990年,第2卷,第465-491页。

③ 刘钝、王扬宗编,《中国科学与科学革命:李约瑟难题及其相关问题研究论著选》,沈阳:辽宁教育出版社,2002年。

也就结束了。30年前,科学史中的内外史论战就结束了;再不需要浪费时间仅做内史或仅做外史了。

不 二 臣

正如我在前一讲中说到的,除了比较同一时期的两种文化之外,比较中国不同时期的同一现象也是有用的。比较的最重要部分就是,一旦发现差异,就要努力解释何以有此差异。如果比较两个不同的时代,这就等于要解释历史变化。在这个例子里,我将讨论一个还没有答案的历史问题。

不二臣是一个影响了科学的中国习俗,这是一个古代惯例,就是禁止一个官员为两个相继的王朝服务。许多学者甚至走得更远,即使他们在旧王朝不曾为官甚或没有为官前景,也拒绝为新王朝服务。

这种禁律似乎没有影响到元朝的汉人。蒙古人任命有才能的汉人天文学家从事始于1276年的授时改历并不困难。这次改历的所有领导者都是忽必烈的私人顾问。他最亲近的顾问刘秉忠,出生于为金朝统治者服务的家族,但是他热情地为忽必烈工作,还招募了那个时代许多擅长所谓"阴阳"——即天文学、占星术和占卜——的高手。1276年南宋政府投降之后,前宋的天文学官员,比如陈鼎,就为蒙古人提供了北方人所没有的科学经验。此前他就参加过一次改历。他选择了效力。他不像前朝人,不愿意死亡。

现在让我们看看明清变迁,这也是从汉人政府到异族政府即满人政府的变迁。许多杰出的学者拒绝了对他们的任命,即使他们在明朝不曾为官。他们成为天文学教师和数学教师,或者成为医生。那个时代最优秀的天文学家梅文鼎、王锡阐和薛凤祚就是例子。王锡阐在清初只有16岁,就试图溺水而亡。有人救了他,但他却以教书谋生度过余生。他一辈子用明代年号来写日期。由于这是重罪,他总是用篆体写作,所以很少有人能阅读他的手迹。

傅山(1607—1684)是一位杰出的行医者和作家,而且还是他那个时代最优秀的书法家和画家之一。在明朝,他从来没做过官。清廷数次任命他为官,并强迫他接受。他称病不去京城,政府就安排把他的床也搬进皇宫,所以他是可以不下床接受官位的。他抵京时宣称,如果逼他接受哪怕是一个名义性的任命,他就自杀。[①] 他的余生极其贫穷。他向平民而不是官员卖药来维持自己和儿子的生活。

我们已看到了宋—元转型和明—清转型之间的巨大差异——不二臣的思想在第一次转变中并不重要,但在第二次转型之中却盛行起来。这种反差的原因是什么,哪种情况影响了科学?我还没有透彻地研究这个问题,因此我没有什么看法。通过历史考虑这个概念的所有维度,我们就能明白。它所对待的,就是艰苦的工作、富有思想的分析、想象和开放的心智。或

---

① 见王锡阐和梅文鼎的传记,金秋鹏主编,《中国科学技术史·人物卷》,北京:科学出版社,1998年,第660-675页。王锡阐的英文传记,见 Sivin, Wang Hsi-shan (1628-1682), in *Science in Ancient China*, 前引书,chapter 5, 1995。傅山的传记,见郝树侯,《傅山传》,太原:山西人民出版社,1981/1985年。薛凤祚没有充实的传记,但可见他在《历学会通》中的作品;有些版本刊载的是同义书名《天学会通》。关于这两个书名的复杂文献目录,见 Sivin, *Science in Ancient China*, 前引书,Chapter 4, pp. 28-29 及注39。

许参加这次讲座的某个人就能解决这个迷人的问题。

结　　语

　　在这次的演讲里我已经给出了一些文化簇的例子。伊斯兰对汉人天文学的影响问题原来不是科学因素问题，而是政治因素问题。另外是比较的例子。李约瑟问题之讨论，比较的是科学革命在不同国家的情况。学者们的广泛兴趣问题，以及拒绝为两个王朝服务的学者的问题，所比较的都是中国史上的不同时期。

　　所有这些例子都只说明了一个简单的观点。尽管训练某种历史分析——例如智识史或社会史——的专家，是大学里的惯例，但是这种狭隘找不到许多历史问题的可靠理解。文化簇的想法有用，简直就像这个提示一样。

　　如果你正被训练成为一个专家，那你如何能够成为一个通才呢？

　　当你开始做研究时，你的老师们会建议你做一个小题目，以便你能够掌握技能。最终，在一个大项目中写一篇学位论文，显示你已经学到了你的专业领域里的技能。从那时起，你就可以决定研究其他专业里的新技能和分析方法了。你可以设计一些项目，考虑一个问题的多个维度。通过扩展你在不同专业里的经验，你最终就能用文化簇去理解那些用狭窄进路不能理解的主题。学术界非常需要你的智力、动机和想象贡献于世界范围的科学史、技术史和医学史。

人名译名表

Aronowitz, Robert A. 罗伯特·阿茹诺维茨
Aristotle 亚里士多德

Bo Shuren 博树人

Chang, Che-chia 张哲嘉
Chao Yuanfang 巢元方
Chen Bangxian 陈邦贤
Chen Ding 陈鼎
Chen Meidong 陈美东
Cixi, the empress dowager 慈禧太后
Cooper, William C. 库伯
Copernicus 哥白尼

Dai Zhen 戴震
Dong Yuyu 董煜宇

Elman, Benjamin A. 本杰明·艾尔曼

Feng Xianliang 冯贤亮
Fu Shan 傅山

Galileo 伽利略
Gao Xi 高晞
Gu Yanwu 顾炎武
Guo Shoujing 郭守敬

Han Qi 韩琦
Hao Shuhou 郝树侯
He Zhiqing 赫治清
Hippocrates 希波克拉底
Hong Huixing 洪辉星
Hong Pimo 洪丕漠
Hong Zhongli 洪中立
Hsu, Francis L. K. 许烺光
Huang Xiaoru 黄小茹
Huang Yinong 黄一农
Hughes, Thomas 托马斯·休斯

Jin Qiupeng 金秋鹏

Keiji, Yamada 山田庆儿
Khubilai 忽必烈
Kleinman, Arthur 凯博文
Kuhn, Thomas 托马斯·库恩

Lackner, Michael 迈克尔·莱克勒
Latour, Bruno 布鲁诺·拉图尔
Levi-Strauss, Claude 克洛德·列维-斯特劳斯
Li Bozhong 李伯重
Li Di 李迪
Li Hongzhang 李鸿章
Li Peishan 李佩珊
Li Shizhen 李时珍

Liao Yuqun 廖育群
Liang Jun 梁峻
Liu Bingzhong 刘秉忠
Liu Dun 刘钝
Liu Yongcheng 刘永成
Lloyd, Geoffrey 罗维(杰夫里·劳埃德)
Lou Zengquan 娄曾泉

Ma Boying 马伯英
Me Wending 梅文鼎
Mencius 孟子
Muller, Herman 赫尔曼·缪勒

Needham, Joseph 李约瑟

Payer, Lynn 林恩·佩耶

Ren Dingcheng 任定成

Shaffer, Simon 西蒙·谢弗
Shapin, Steven 史蒂文·夏平
Sheng Kuo 沈括
Shigehisa, Kuriyama 栗山茂久
Sivin, Nathan 席文
Su Shi 苏轼(苏东坡)
Su Song 苏颂
Sun Simiao 孙思邈
Sun Xiaochun 孙小淳

Tang Tingxian 唐廷献
Tao Yufeng 陶御风
Tongzhi, the emperor 同治皇帝

Traweek, Sharon 沙伦·特拉维克

Vesalius 维萨留斯
von Bel, Johann Schall 汤若望

Wang Xishan (Xichan) 王锡阐
Wang Yangzong 王扬宗
Westman, Robert S. 罗伯特·韦斯特曼
Woolgar, Steve 史蒂夫·伍尔加

Xi Zezong 席泽宗
Xu Guangqi 徐光启
Xue Fengzuo 薛凤祚
Xue Pangao 薛攀皋

Yabuuti, Kiyosi 薮内清
Yan Su 燕肃
Yano, Michio 矢野道雄

Zeng Guofan 曾国藩
Zeng Xiongsheng 曾雄生
Zhang Heng 张衡
Zhang Zhidong 张之洞
Zhang Yinlin 张荫麟
Zheng Shu 郑术
Zhu Bangxian 朱邦贤
Zhu Kezhen 竺可桢
Zhuang Tianquan 庄添全
Zuo Zongtang 左宗棠

译 后 记

2009年3月起,作为科学人文领域的初学者,我在北京大学任定成小组科学文化史讨论班上与大家一起研读罗维和席文两位教授合著的《道与名》(*The Way and the Word: Science and Medicine in Early China and Greece*)一书。结合阅读,我也试着翻译此书。就在这时,恰巧席文教授应中国科学院自然科学史研究所之邀,担任当年的竺可桢讲席教授,于2009年4月13日至22日在北京大学做了主题为"科学史方法论"的系列讲座。我和百余位来自北大、清华、北师大、人民大学、中国科学院自然科学史研究所的同学们一起聆听了每一次讲座。讲座结束后,席文教授于4月23日上午来到承泽园,参加了任定成小组的讨论班。本期讨论班的主题是"'文化簇'与中国科学技术史研究"。在讨论班上,我简略介绍了我对"文化簇"的理解,得到了席文教授的当面指教。①

席文先生的演讲为我们展示了研究科学史的新进路和新方法,对于我们了解科学史的新变化,对于我们从新的视角选择研究主题、开展研究,都很有意义。任定成教授和孙小淳研究员积极推进此系列演讲的汉译工作,北京大学出版社周雁

---

① 任定成等,"'文化簇'与中国科学技术史研究——北京大学'科学文化史'讨论班述要",《中国科技史杂志》第31卷(2010年)第1期,第14-25页。

翎编审对此做出积极响应。在老师和同学们的鼓励下,我接受了翻译这个系列演讲的任务。

我的译稿初稿完成后,在任定成小组讨论班上逐章汇报过,小组成员的批评和建议提高了我对演讲稿内容和相关背景知识的理解。译稿经任定成教授校订、席文教授审定,先后发表于《北京大学学报(哲学社会科学版)》、《清华大学学报(哲学社会科学版)》、《南开学报(哲学社会科学版)》、《浙江大学学报(人文社会科学版)》和《复旦学报(社会科学版)》。这些刊物的编辑根据同行专家的匿名审稿意见,对译文提出了一些修改建议。译文发表后,任定成教授在中国科学院研究生院"科技史理论与方法"讨论课上组织选课同学讨论了这些演讲,我在讨论中受益匪浅。这次结集出版,我又改正了少量翻译错误。这本书虽然是由我译出的,但却吸纳了老师和同学们的思考和智慧。

借此书出版的机会,我感谢席文教授的指教,感谢北京大学任定成小组讨论班主要成员对原文理解提出的不同见解,感谢中国科学院研究生院选修"科技史理论与方法"讨论课的同学们对译文所做的评论,感谢上述5种大学学报的编辑和审稿者对提高译文质量作出的贡献,感谢北京大学出版社的编辑付出的劳动。

虽然在翻译过程中我已经尽了力,但多次更正翻译错误的经历告诉我,现在呈现在读者面前的译文仍然可能有我尚未察觉到的错误。我期待着本书读者的指教。

任安波

2011年11月8日

于北京大学承泽园

Preface

It was a great honor to be asked by the Chinese Academy of Sciences to deliver the Zhu Kezhen lectures in 2009. They were given at Beijing University, to encourage attendance by students, between 13 and 22 April. [1]

Because I was unable to visit China until 1977, I never met Prof. Zhu Kezhen, who lived from 1890 to 1974. But I knew his work, because his writings were essential to every student of the history of Chinese science. He wrote on a very wide range of historical issues, from geology to meteorology to astronomy. The study of Shen Kuo that he published in 1926 was one of the inspirations for my own work on Shen. His article entitled "Which should we study the history of ancient science in our country?," which appeared in *Renmin Ribao* (*People's Daily*) in 1954, inspired many readers. [2] In short, he was the first of many modern Chinese scholars from whom I learned the art of historical research.

---

[1] I acknowledge with gratitude the generous help of Prof. Ren Dingcheng and the members of his Center for Social Studies of Science at Peking University, and Prof. Sun Xiaochun of the Institute for the History of Natural Sciences at Chinese Academy of Sciences in the preparation and delivery of the lectures and their preparation for publication. I am also grateful to Zheng Shu of the Institute for the History of Natural Sciences for her aid and advice.

[2] Zhu Kezhen 1926. "Contributions and records of geology by Shen Kuo of the Northern Song period," *Kexue*(*Science*), 1926, 11(6), 792-807; Zhu Kezhen, "Which should we study the history of ancient science in our country?," *Renmin Ribao* (*People's Daily*), August 27, 1954.

That is why I have devoted these lectures to methods of historical research. I have spent fifty years on reading, thinking, and writing about the history of Chinese science.① During that period, methods of general historical investigation have changed in fundamental ways. Most of these changes took place as historians adapted the tools of other disciplines to their own needs. Half a century ago, most of what was called intellectual history relied on the methods of philosophical analysis. Over the 1960's and 1970's, demographics turned out to be useful, then economics, sociology and anthropology. Over the past decade or so, the methods of environmental studies have led to environmental history.

The use of social science gradually changed the emphasis of leading historians from intellectual history to social history before 1970. By 1980, many of my colleagues in general history realized that social history, like intellectual history, is too narrow to encompass the complexity of history. The result was a growing interest in looking at problems from several viewpoints at the same time. Cultural history, although those who study it do not agree on its definition, tends to look at the arts, material culture, economics, and social institutions of a historical period. The more recent approach called "cultural manifolds," described in the fifth lecture, goes a little further, first examining a problem to determine all the pertinent dimensions, and then exploring all of them as part of a single pattern. Other equally important recent studies have encouraged scholars to a-

---

① Here and in the chapters that follow I use "history of science" to include the history of technology and medicine as well. This book, because of space limits, will not discuss the history of technology.

void assuming that all members of the group they study agree, and to recognize that social decisions are negotiated outcomes of disagreement and conflict among diverse individuals.

These new developments have changed general history in striking ways. They have also affected the most enterprising historians of science and, gradually, their pupils. Partly because of these new approaches, historians of physical science are more likely than before to investigate such problems as scientific fraud, plagiarism, conflicts with the public interest, the role of the public and of the government in the funding of science, and the self-promotion of scientists. Historians of technology no longer see the techniques of the past as applied science; craftsmen before the 19th century were seldom trained in the science of their time, and passed down unique skills.

In many countries, the history of medicine is divided. Some historians, mainly those who teach in medical schools, still see their field as a chronicle of progress, and pay little attention to therapy by non-physicians. Others, mainly those in departments of history and the history of science, are more apt to explore the much wider sphere of public health, and to examine the roles of self-interest and social status in the relations of doctors, patients, and the state. Because work of a more traditional kind continues, the outcome is more balanced, more diverse research.

Probably the largest change has been the growth in studies of recent and contemporary history. In the 1950's, the history of science in the West was mainly concerned with the beginnings of science and with the origins of modern science in the Scientific Revolution. The scientists and physicians who did such research had been educated at the beginning of the 20th century. But now most historians of science are educated in departments of history or of the history of science. They seldom know Greek and Latin, so they are unable to work on any period before the mid 19th century, when technical teaching in Latin began dying out.

*　　　*　　　*

The purpose of these lectures is to encourage young historians of science to think about the broadest range of methods that they are able to learn about and use. There are five topics:

"How the History of Science and Medicine is Changing" discusses how the changes in general history summarized above have affected research on science and medicine. It examines some new approaches that are currently being used in China and elsewhere. It also gives some examples of new tools of investigation that could be profitably applied, and some special characteristics of early Chinese science that affect how research is done.

"Using the Methods of Sociology and Anthropology" explains the advantages of using these approaches. Both originated as methods for the study of contemporary societies, but scholars have found them greatly useful for understanding every stage of history, from the most ancient to the most recent. In medicine they facilitate recognizing changes in the meaning of technical language and in understanding the many kinds of therapy done by people who are not physicians. They provide ways of thinking about the effectiveness of medicine in ways broad enough to cast light on ancient health care.

"Using Studies of Popular Culture" discusses the methods most valuable for studying the culture that all of China's people shared. Most of the studies of the history of science have been concerned with the work of famous members of the elite. A few historians have tried to provide balance by discussing the very different approaches of ordinary people, most of whom until recently were not educated and could not read and write. Using medicine as an example, this chapter discusses a variety of recent methods helpful in such work.

"Using Comparison" examines a kind of historical work that can be enlightening when done with imagination. Comparison is useful in leading to original understanding not only with respect to two different cultures, but with respect to the same culture at two different times. In other words, it enables a better analysis of change, which after all is the point of historic research. The chapter discusses the importance of accurately comparing the meanings of words in different languages, and takes up some examples of problems that can only be solved by careful comparison.

"Using Cultural Manifolds" is about overcoming the limitations of historical specialization in order to make sense of topics that require a multi-disciplinary point of view. The basic idea of specialized research is to explore questions deeply and rigorously. The originators of this idea, in the 19th-century German universities, believed that generalists would combine the results of such narrow work, step by step, to provide valid and comprehensive results. In the modern world, such step-by-step synthesis has failed to take place. Most writing on the history of science is designed for fellow specialists. Scholars' understanding of other fields is frequently twenty years out of date. Books designed for a popular readership are usually not comprehensive enough, and not rigorous enough, to build a coherent structure of learning. The use of cultural manifolds allows scholars doing research in primary sources to attain results broad enough to be generally interesting and comprehensive enough to be valid.

In studying China before modern times, I have learned a great deal from early authors who wrote about nature and man's relation to it. I suspect that people such as Zhang Heng, Shen Kuo, Su Song, and Li Shizhen would have been enthusiastic about the powerful results of work that scientists do today. But at the same time, they would have found modern science profoundly unsatisfactory in two important respects. First, early Chinese believed that the purpose of such study is to throw light on the Way (*Dao*). Because the Way is a moral and esthetic concept as well as a cosmological one, a correct understanding of the natural world must account for justice and beauty as well as physical reality. The idea of science that is objective, with no moral or esthetic meaning at all, that has nothing to say about how one ought to live, would have been repugnant to them. They would perhaps be more aware than modern people are that a science independent of such values is bound to lead to an environmentally polluted world, and to scientists who care no more about justice than people with no education. Although ancient scholars would respect the accuracy of modern science, I suspect they would find it much too narrow. Fully comprehending their ideals of knowledge may help us to widen it.

The most inspiring sentence I have read comes from the *Lunyu* (*Analects*): "When you know a thing, to recognise that you know it, and when you do not know a thing, to recognise that you do not know it. That is knowledge."[①] After many years I came to realize that Confucius was not writing about knowledge of physical laws or theories, but about how to live.

Nathan Sivin
Chestnut Hill
22 September, 2011

---

① *The Analects of Confucius*, translated and annotated by Arthur Waley, New York: Vintage Books, 1989, 2.17.

How the History of Science and Medicine is Changing

N. Sivin

2009. 4. 13

   In the 1950's when history of science and medicine become professional fields, historians of science and medicine, who were trained in science and medicine, investigated thoughts of great men by the tools of philology and evidential research. By about 1970, most of historians of science and medicine were being educated in history, and began to write for laymen. Since 1980's, many scholars have moved away from research on ancient times and studied the recent past, took the approaches of sociology and anthropology to un-famous scientists, administrators, patients and their families, and analyzed social status, relations between people, the cost of research, and the role of authority. Recently, the focus has moved to local cultures of modern science. It would be worth while to study the widespread involvement in science of literati during the Northern Song, the costs of scientific research before modern times, links between science and literature, and the effectiveness of therapies, by combining intellectual and social history, drawing on recent social science, and developing a good balance between conventional and innovative approaches.

1 How the History of Science and Medicine is Changing

Many people think of the history of science and medicine as two fields.① That is because sixty years ago scientists wrote on the history of natural science, and physicians wrote on the history of medicine. Keep in mind that Professor Zhu Kezhen, one of the first great modern historians of Chinese science, was a meteorologist by training and profession. Most of the most important historians of science and mathematics in the sixty years after Professor Zhu began working, such as Professors Xi Zezong and Chen Meidong, were also trained in science. Historians of medicine were educated in schools of either Chinese or Western medicine. History of medicine and history of science did not originally have much contact, as we can see in the separation of the Institute for the History of Natural Sciences in the Chinese Academy of Sciences, and the Zhongguo Yishi Wenxian Yanjiusuo in the Zhongyi Yanjiuyuan.

Today, because the two fields have influenced each other so much, many of their research questions and methods are not greatly different. The fact that Professor Liao Yuqun, a historian of medicine, is now Director of the Ziran Kexueshi Yanjiusuo shows that the separation is less important than it used to be. Both institutes train their own graduate students to

---

① This lecture is respectfully dedicated to the memory of Professor Zhu Kezhen, one of the few scholars who introduced the modern history of science into China.

be professional historians. In both, as in universities abroad, research questions and methods have been changing quickly. Today I will talk about what some of those changes have been, and what kinds of change are continuing. I will mainly look at changes that are similar in the history of science and medicine rather than those that are different in the two fields. For that reason, I will often use the word "science" for both.

Professional History of Science

I will begin with the 1950's, since that was when the history of science and medicine became professional fields. When I say "professional field," I do not mean the beginning of high-quality scholarship. Competent research in both fields is centuries older than that in both Asia and Europe. It became professional when people could make a living as researchers, paid to teach in university departments or to work in research institutes.

About 1950, many scholars were working in the history of science or medicine. Almost all were still trained in science, not in history. What they wrote about was, in the past, how ideas in their field had developed to resemble those of modern science. In their eyes, anything not recognizable by modern criteria was not worth studying. They studied their sources critically, using the tools of philology—that is, critically reading and analyzing what texts said. Europeans had their tradition of philology, and Chinese, Japanese, and Koreans used the similar and equally versatile tools of evidential research (*kaozheng*). Both traditions were extremely powerful for learning who wrote a document, whether it was authentic, how it was related to other documents, and what the words in it meant.

The resulting histories of science and medicine were usually summaries of concepts and methods, as they appeared in important books, in chronological order. "Change" ordinarily meant differences in the contents of two books. The purpose of that kind of research was to identify "accomplishments (*chengjiu*)"—who did something first, and who did it better. It portrayed the small number of Great Men that historians studied as heroes, unlike ordinary people. Copernicus, Vesalius, Shen Kuo (or Gua, 1031 – 1095), and Guo Shoujing (1231 – 1316) had very little in common with other people they worked with. Such histories were not able to explain the reasons for this difference. Nor could they account for change, except by vague explanations such as influence. They could not say why sometimes influence was accepted and sometimes it was rejected. In trying to understand the relations between Great Men, they looked primarily at organizations. [①]

This approach was successful for a long time. Its audience, like its authors, were scientists and physicians. They wanted to know who their intellectual ancestors were. They believed that the knowledge of their own time, as they had learned it from textbooks as students, was highly reliable and rational. Their own work was meant to improve the

---

① See the discussion in Huang Xiaoru 2008.

concepts and methods of their own time. They saw history as a matter of tracing that same pattern of improvement backward. Their history of science had very little in common with general history, which is about the endlessly complicated, as often irrational as rational, experiences of human beings in societies.

Developments in General History of Science

By about 1970, most historians of science and medicine in the West were being educated in history rather than in science. They began to write for laymen rather than scientists. They had learned from the Vietnam War (1959 – 1975) that science and technology were used often to kill as well as to save life, and that they could be a tool to destroy the environment as well as to improve it. Historians' concern about the political misuse of science led many of them to move away from ancient times and study the very recent past. Increasingly, graduate students chose to study contemporary science. For instance, of the twelve professors in my own department, ten study the twentieth and even the twenty-first century, and very few of our graduate students study events before 1900.

When you study your own time, you are bound to notice that economic necessities of scientists, their political decisions, their competition, and their personal relations are too important to ignore. You notice that scientists with average training, who will never be famous, do most of the work, and produce much of the change. Investigating all of these matters, and many others, is a routine part of research today.

That raises a question that, once you begin asking it, seems obvious: isn't the same thing true of the past? If so, a history based mostly on the thoughts of Great Men is too narrow to explain how science really evolved.

By about 1970, while many historians of science such as myself were asking ourselves such questions, we found that anthropology and sociology were evolving research methods that could be very helpful when applied to the past. In the next lecture I will be dealing with the influence of these disciplines, so here I will just say a few words about it.

The old history of science was about the ideas and theories of heroic individual scientists. That can lead to valuable conclusions, but not to balanced ones. To restore the balance, it became essential to study the role of society and culture in shaping technical change. All scientists and physicians today are aware of differences in social status, in conventional and unconventional relations between people, in wealth, in authority, and so on. Those are what sociology studies. Scientists and physicians also learn from people around them how to understand and classify their experience, how to express themselves in ways that will persuade other people, and so on. Those perceptions and methods that people share are what we call culture, and they belong to anthropology. Most social scientists in both disciplines investigate the present, not the past.

Social history of science began in the 1950s as nothing more than the study of scientific organizations: the Royal Society in England, the Academy of Sciences in France.

Because those organizations admitted only a few scientists, who were usually outstanding rather than typical, this research did not lead to many important innovations in understanding. Since then, as historians have applied anthropology and sociology to the past, they have thrown light on every aspect of technical careers. Why, in Europe, could people not be employed as physical scientists or mathematicians until the 18th century, although in China scientists have been bureaucratic officials for the past two thousand years? Why, in China before the 20th century, did scientists seldom disagree in public, and even more seldom argue publicly with living rivals? Why did astronomical officials disagree more often than mathematical officials? From the Greeks on, scientific argument and public face-to-face debate were normal in Europe. What accounts for this obviously important contrast? These differences point to deeper differences in social conventions, priorities, and values—in other words, to differences in society and culture.

Let me give some specific examples of important changes in thinking historically about technical change, first in Europe and then in China.

First, new ideas do not automatically convince people that they are correct and better than old ideas. Someone has to invent not only new technical methods but also new means of persuasion. If the ideas are revolutionary, scientists must also invent a new public to persuade. Without these social inventions, change can be extremely slow. Although Copernicus was widely respected as an astronomer, Robert Westman (1980) showed that, in the three generations between 1543, when his book was published, and

1600, only a total of ten people in Europe accepted Copernicus' theory that the earth was a planet, and that all the planets rotated about the sun. Copernicus was not a revolutionary. Because he wrote for the conservative scholars of the universities, new understanding came too slowly until Galileo invented a new public outside the universities for him.

In seventeenth-century England, universities mainly prepared people for religious careers. It is not surprising that most of the important scientific discoveries of the time were neither discovered nor developed in universities. Since universities were not prepared to approve them, who would judge and accept the work of innovators? Steven Shapin and Simon Shaffer, in 1985, produced an important answer to this question, which few people before them had even asked. It was gentlemen, educated men with high social standing and an interest in science, who demonstrated their discoveries and hypotheses to each other. At first they did so in their own houses for a few visitors. In 1660, by they organized the Royal Society,① the first European scientific society. In this organization they could show and discuss their new work with a much larger group of gentlemen, and publish it in the Society's journal as true science. This is the pattern that has become normal today in such journals as *Studies in the History of Natural Science*, but in the 1660s it was a new invention.

Another example is the invention of the standard fruit fly

---

① Its full title is "The Royal Society of London for the Improvement of Natural Knowledge."

(*Drosophila melanogaster*; Kohler 1994). In the early 20th century, Herman Muller and his colleagues in the biology laboratory at Columbia University began breeding a special fly of this species that was extremely convenient for experimenting on problems of genetics. Over a generation, they gave these insects to many laboratories doing the same work. The result was that the special fruit flies became the standard insect for many kinds of research. These laboratories depended on them, and repaid these gifts by informing the group at Columbia University about the progress of their work, and often accepting their advice. As a result, the group at Columbia became, and remained for a long time, dominant in that kind of experimentation. Thus an insect can become a tool of domination in experimental science.

In both of these examples, we can see that the ideas and theories of science depend on social activities—the invention of new relationships among the new scientists, even the invention of a new kind of insect—to make possible their development in certain directions. That makes it possible to answer the question of why an astronomical reform in China employed more than 150 specialists in 1280, but no such project in Europe could bring together more than five or six astronomers for centuries after that. The answer lies in China's centralized bureaucratic government.

In the history of medicine, there has been an equally important shift over the past 20 years. Almost all study before that time was about physicians and their practice. Doctors' writing, still the main source for medical history in the 1980's, did not record in detail the experience of patients. But there were a great many such records in biographies and diaries. Since that time we have had widely read books on the experiences of patients, and are beginning to understand the history of pain in ordinary life, in disease, and in surgery (Porter & Porter 1988, 1989).

Chinese medical books also give very little information about patients and their experience. But a decade ago, the doctoral dissertation of Zhang Zhejia (1998) used the documents in the Beijing Imperial Palace Archive to trace in remarkable detail the experience of the emperor Tongzhi (1861 – 75) and the empress dowager Cixi (in that position 1874 – 1908) during their treatment by medical officials and unofficial doctors. They were not typical patients, but Zhang's work opens the way for studies of more typical ones.

A Change of Focus

It was once usual to consider modern science as one body of knowledge, the same everywhere. Since its concepts and language are supposed to be universal, scholars assumed that scientists from different cultures think and act in the same way. But are the mental habits of chemists in very two different cultures really the same? A recent study suggests that translating scientific writing from one language to another changes its meaning in ways that are too important to ignore (Montgomery 2000). This was true in early times, the author concluded, and continues to be true today. It would be worth while to do such research on differences in content and thought between technical papers on the same topic written, for instance, by Chinese and Japanese scientists.

Again, studies of scientific concepts alone are too narrow to answer broad questions. Sharon Traweek did a remarkable field study in which she compared the linear accelerators (*zhixian jiasuqi*) at Tokyo and Stanford Universities and the physicists who used them. Hundreds of physicists, organized in research teams, needed to use these machines. Those in charge of them assigned blocks of time for each experiment. Any project that could not get enough time would not succeed. Traweek found that the way decisions were made about assigning time were quite different in California and Tokyo. The difference did not depend on the physics, but on the administrative practices, the working habits, and the human relations of each place. A comparative study of the Japanese and American rocket projects gave additional evidence for this conclusion (Sato 2005). To understand the success or failure of scientific research projects, in other words, one has to recognize that there are local cultures of modern science, and study them.

Work on this kind of problem has not progressed very far, particularly on the scientific cultures of China. In *every* country involved intensively in scientific research, plagiarism is a problem. A number of studies have investigated the many cases in the United States in recent years (Judson 2004, chapter 7). In China, a published investigation suggests that plagiarism deserves historic study (Li & Xue 1996), but after another dozen years we do not know more than that. As we know, copying was a widely accepted practice in Europe before the 20th century. But no one has studied it in China. Some scholars may think of plagiarism as shameful, but it has been a fairly common practice in science from very early times, and impartial studies of it are essential if we want to understand science as a historic phenomenon.

New Initiatives in the Study of Chinese Science

Historians of Chinese science in Asia have begun applying a number of these new methods and approaches to their own work.

The Japanese scholars Yamada Keiji and Kuriyama Shigehisa have combined intellectual and social history in a variety of books on medicine and science. For instance, Yamada's study of the *Shoushi li* (1980) was the first book to examine the social, political, and institutional aspects, at the same time as the technical aspects, of a Chinese astronomical reform. I found it extremely useful when writing my own book on the same reform (2008). Ma Boying's cultural history of Chinese medicine (1994) is a very rich survey of the many dimensions of its topic. Feng Xianliang (2002) is one of several scholars in China who have been using new methods to study environmental history. As young scholars are trained to choose from a wide variety of research techniques, China will no doubt develop a good balance between conventional and innovative approaches.

New Tools to Apply

Let me list a few areas in Chinese scientific history about which we know little or nothing, and tools that will be useful in studying them.

• Years ago, in a biography of Shen Kuo, I noted that many men in the Northern Song period were exceptionally skilled in almost all the arts and sciences of the time, from painting and poetry to mapmaking, invention, mathematics, astronomy and alchemy. Such scholars were rare in the centuries before and after. I am surprised, after thirty years, to find that no one has yet explained this interesting phenomenon. No doubt a number of scholars have opinions about it, but this is a research question, and the research has not yet been done. Surely this issue, patterns of intellectual scope, is important enough for someone with broad interests to study.

• Very little research has been done on the costs of scientific research before modern times. There has also not been much study of the large trade in medicines, either within the empire or international. Economic history provides techniques for dealing with just such questions. For instance, Prof. Li Bozhong of Beijing University has applied economic techniques to the history of agriculture and industrialization (2000, 2007). We know from studies of old Chinese drug stores that there are rich records that scholars who know quantitative economics can analyze (Liu

Yongcheng & He Zhiqing 1983, Tang Tingxian 2001). Dong Yuyu's recent investigation of the Northern Song government's monopoly on the sales of almanacs (*liri de zhuanmai*, 2007) provides a model.

● The study of science and literature in Europe has developed considerably in the last generation, and offers many fruitful topics of investigation. This work has hardly begun in China, although its ancient scientists and physicians usually had greater skill in poetry and other arts than their counterparts in the West. Their collections of literature and poetry often survive, and are often large. Everyone who loves poetry knows that it expressed feelings that were difficult to write about in prose. Studying it offers access to thoughts and feelings that scholarship would otherwise overlook, and of course will clarify the important links between literature and science in early China.

● Probably the most neglected area in the history of medicine is the effectiveness of its therapies, not only drugs but its many other types of treatment. Historians have tended either to accept every statement from an early physician about successfully curing patients, or else to reject most of them. Neither is a rational approach. How can we evaluate curative claims? From a narrowly technical modern point of view, laboratory testing has not yielded an answer.

We have learned from medical anthropology that therapy in a modern laboratory cannot produce results identical to what the same drugs or the same manipulation will produce in the original social and cultural circumstances. To understand results, in addition to techniques we have to understand those social and cultural circumstances. We also need to study the actual relationships between therapists and patients, the character of their interaction, the role of the family or others who were present, and so on. My present research project deals with the question of efficacy a thousand years ago, because a study of all the dimensions of therapy is the only reliable way to understand how health care developed before modern times. Such a study will reward exploration by many scholars, the more the better.

Further Steps

In a book published five years ago, my colleague Geoffrey Lloyd and I argued that, in order to draw the most reliable conclusions, studying all the dimensions of a problem is advisable, and comparative studies are often useful. Comparison is actually a broad method of investigation. It does not necessarily require comparing different two civilizations. We can also explore two times or places in the same society.

Historians tend to study only the history of ideas or only social history, but those are artificial distinctions based on European habits. In studying ancient Chinese scientists, it is also necessary to comprehend the official bureaucratic culture to which most of them belonged, the ways scientists persuaded each other, and so on. I suggested that the most productive approach is to first ask which dimensions—personal, political, social, economic, organizational, artistic, and others—are pertinent to a given problem. Then we can examine all of those relevant dimensions. That lets us understand not only how each one bears on the problem, but how they interact. I will be dealing with comparison and cultural manifolds in later lectures in this series.

In closing, I just want to leave you with the idea that it is possible to considerably expand the scope of scholarship on Chinese science. Professor Zhu Kezhen ended an essay he wrote 65 years ago with these words:

"If I can interest such a learned group in looking into these problems, this essay will not be wasted effort."[①]

References

Chang, Che-chia. 1998. "The Therapeutic Tug of War. The Imperial Physician-Patient Relationship in the Era of Empress Dowager Cixi (1874 – 1908)." Ph. D. dissertation, Asian and Middle Eastern Studies, University of Pennsylvania.

Dong Yuyu. 2006. Cong wenhua zhengti gainian shenshi Song dai de tianwenxue—yi Song dai de liri zhuanmai we ge'an (An examination of the astronomy of the Sung period using the concept of cultural manifold: the case of the Sung monopoly on the sale of almanacs). In Sun Xiaochun & Zeng Xiongsheng, editors. 2007. *Song dai guojia wenhua zhong de kexue* (Science and the State in the Song Dynasty), pp. 50 – 63. Zhongguo kexue jishu Chubanshe.

---

① 竺可桢,二十八宿之时代与地点, in 竺可桢文集 (Kexue Chubanshe, 1979), 234 – 54, citing p. 253.

Feng Xianliang. 2002. *Ming Qing jiang nan di qu de huan jing bian dong yu she hui kong zhi* (Environmental change and social control in Ming and Qing Jiangnan). Xueshu chuangxin, Shanghai Renmin Chubanshe.

Huang Xiaoru. 2008. Zhongguo jinxiandai kexueshi yanjiu zhong de tizhihua wenti chuyi (A Brief Review of Institutionalization as an Issue in Research on the History of Modern and Contemporary Science in China). *Zhongguo kejishi zazhi*, 29. 1: 30 - 41.

Judson, Horace Freeland. 2004. *The Great Betrayal. Fraud in Science*. New York: Harcourt.

Kohler, Robert E. 1994. *Lords of the Fly: Drosophila Genetics and the Experimental Life*. Chicago University Press: Chicago.

Kuriyama, Shigehisa. 1999. *The Expressiveness of the body and the Divergence of Greek and Chinese Medicine*. New York: Zone Books.

Li Bozhong. 2000. *Jiangnan de zaoqi gongyehua*: 1550 - 1850 *nian*: 1550 - 1850 年. Beijing: Shehui Kexue Wenxian Chubanshe.

Li Bozhong. 2007. *Jiangnan nongye de fazhan*, 1620 - 1850, 1620 - 1850. Shehui jingji guannian shi cong shu. Shanghai Guji Chubanshe. Translation from Li, *Agricultural development in Jiangnan*, 1620 - 1850 (Studies on the Chinese economy; New York: St. Martin's Press, 1998).

Li Peishan; Xue Pangao. 1996. Shi Yingwen wenti, haishi kexue diode wenti? (Is the problem English or scientific ethics?). *Tzu-jan pien-cheng-fa t'ung-hsun*, 4: 74 - 80. For a report on this study in English see *Science*, 1996. 10. 18, 274: 337 - 338.

Liu Yongcheng & He Zhiqing. 1983. *Wanquantang de youlai yu fazhan* (Origins and development of the Hall of Complete Health drugstore). Zhongguo shehui jingji shi yanjiu, 1: 1 - 15.

Ma Boying. 1994. *Zhongguo yixue wenhua shi* (A history of medicine in Chinese culture). Shanghai Renmin Chubanshe.

Montgomery, Scott L. 2000. *Science in Translation : Movements of Knowledge through Cultures and Time*. Chicago: University of Chicago Press.

Porter, Roy, & Dorothy Porter. 1988. *In Sickness and in Health: the British Experience*, 1650-1850. London: Fourth Estate.

Porter, Dorothy, & Roy Porter. 1989. *Patient's Progress: Doctors and Doctoring in Eighteenth-century England*. Cambridge, UK: Polity Press.

Sato, Yasushi. 2005. Local Engineering in the Early American and Japanese Space Programs: Human Qualities in Grand System Building. Ph. D. dissertation, History and Sociology of Science, University of Pennsylvania.

Shapin, Steven, & Simon Schaffer. 1985. *Leviathan and the Air-pump: Hobbes, Boyle, and the Experimental Life*. Princeton University Press.

Sivin, Nathan. 2008. *Granting the Seasons: The Chinese Astronomical Reform of 1280, With a Study of its Many Dimensions and a Translation of its Record*s. Sources and Studies in the History of Mathematics and Physical Sciences. Secaucus, NJ: Springer.

Tang Tingxian. 2001. *Zhongguo yao ye shi* (History of the Chinese pharmaceutical industry). B: Zhongyi Keji Chubanshe.

Traweek, Sharon. 1988. *Beamtimes and Lifetimes: the World of High Energy Physicists*. Harvard University Press. Comparative study of Tokyo and Stanford linear accelerators and the particle physicists who used them.

Westman, Robert S. 1980. The Astronomer's Role in the Sixteenth Century: A Preliminary Study. *History of Science*, 18: 105-147.

Yamada Keiji. 1980. *Jujireki no michi. Chūgoku chūsei no kagaku to kokka* (The road to the Season-Granting system. Science and the state in medieval China). Tokyo: Misuzu Shobō.

Using the Methods of Sociology and Anthropology

N. Sivin
2009. 4. 14

Anthropologists' aim is to discover shared structures of meaning, which differed from one people or even one group to another. All parts of the shared sense of reality are called culture. People in two cultures do not have exactly the same diseases. There are many ways to divide all physical abnormalities into diseases, and people in different cultures make different choices. Belief can cause and also overcome real physical illness, for the body responds to belief and meaning. Sociologists' aim is to comprehend the structures, norms, and values of the society in which individuals live. Sociology can aid in understanding differences in the authority of doctors in different societies; authority is an important factor in their ability to cure illness. New diseases, and changes in the names of diseases, are not only based on improvements in research. Physicians, like all other members of society, respond to values that are widespread in their society. If they do not, they cannot be very successful in curing sick people.

My topic is anthropology and sociology, and most of my examples will come from the history of medicine.① That is because the sociology of medicine and medical anthropology are well developed, and offer many good examples. But the social sciences can also be useful for studies of science and technology.

To begin with, I will assume that you have not studied these fields. I will define them and explain their use in the kinds of history we do. Both fields are very old, and began to take on their modern shapes in the 19th century. At first, the two were clearly different in their aims and methods.

Most of the famous anthropological studies in the 20th century involved non-Western peoples, such as Africans and Pacific islanders, often in colonies of the European powers. Their aim was the study of culture, a word that has many meanings. In anthropology over the past century, it has come to mean the patterns people shares, and uses to classify and understand their experience. For instance, in Chinese culture people think of brown as dark yellow, but in U.S. culture brown and yellow are not related. Neither culture is wrong; it is simply a cultural difference.

Sociology, on the other hand, began when scholars began

---

① This lecture is respectfully dedicated to the memory of Professor Xi Zezong.

analyzing the structures of their own societies. They explored differences in social status, differences in wealth, differences in authority, and so on. These differences vary in every country. For instance, in the U. S. and China, relations between professors and students are very different. The ways two people relate to each other depends on these differences.

I will discuss anthropology first.

Anthropology

Anthropology has several branches. The one I will be talking about today is usually called cultural anthropology in the United States and social anthropology in England.[①] This kind of scholarship became important when it developed the technique called participant observation. Anthropologists lived among people unlike themselves, taking part in their work and observing them. They studied their daily lives and their special periodic rituals (*liyi*, *lisu*). Their aim was to discover shared structures of meaning, which differed from one people or even one group to another. How did people think of themselves and other people? How did people interact? How did people understand and classify their own society and the non-human world? How did people understand change? How did they explain, for instance, health, illness and recovery?

[①] These two specialties are not identical, but the distinction is not important here. This lecture is not meant as a general introduction to the social sciences. It simply summarizes those aspects of both that I have found particularly useful in the study of scientific history.

2 Using the Methods of Sociology and Anthropology   129

These are all parts of the shared sense of reality that anthropologists called culture. The aim of most anthropologists has been to understand the patterns of culture. ①

It is difficult to talk about ritual in the Chinese language. Because not many people study anthropology, many students think of ritual as unscientific and therefore not worth studying. But personal relations are not scientific, and if we want to understand them, we use concepts that let us analyze their complexity. Anthropology and sociology offer a number of highly sophisticated concepts and methods that have proved helpful in studying many cultures.

One of these is the idea of ritual. This concept is much broader than the ancient Chinese idea of *liyi*. It refers to all patterned behavior that is effective and also has meaning beyond its effectiveness. Everyone in this room understands that, after my host introduced me today, I an expected to speak and everyone else is expected to listen. The host's introduction effectively gives information about me, but its meaning also prepares everyone to listen to a lecture instead of everyone talking. Every student learns to understand this ritual of introduction. That is why it is effective, and why its meaning gives the speaker a certain authority. Ordinary life is full of rituals. They reveal basic patterns of culture.

Let me give an example of ritual's usefulness. One of the

---

① Anthropology is such a diverse discipline that it is impossible to find a single book that describes it adequately for beginners. To get a sense of its current—and always changing—state, I recommend reading recent issues of such journals as *American Anthropologist*, *Anthropology & Medicine*, *Journal of the Royal Anthropological Institute*, and *Medical Anthropology Quarterly*.

many interesting topics in anthropology is the change in a person's life from one status to another. In most societies, the main changes are birth, the transition from childhood to adulthood, the change from being single to being married and forming a family, and death. Generally, for each one of these changes there is a ritual or ceremony that guides the person out of an old status and into a new one: the ceremonies of birth, of adulthood, of marriage, of the birth of children, and of death. The marriage celebration uses symbols and special kinds of behavior to teach a young man and woman how to be husband and wife, to prepare other people to accept them as husband and wife, to register them as legally married, and so on.

In China the complicated details of these ceremonies for the ancient ruling classes were recorded in *ritual classics*. In China today, the ceremonies for these transitions vary from city to countryside, from one level of society to another, from one part of the country to another, from rich to poor, and from one ethnic group to another. But everyone passes through the same stages of life, and their subculture teaches them how to celebrate and mourn.

Scholars form smaller cultures of their own. An anthropologist who attends a meeting of the Chinese Society for the History of Science can analyze its culture by learning what the officers do, who talks in meetings, how the papers are chosen, who decides to attend the presentation of each paper, who speaks first in a group, how people talk to each other in the hallways, and so on.

An anthropologist who wants to understand the culture of physics can look at how people behave when someone receives the

Ph. D. degree, when someone begins work in a research institute or university department, when someone becomes an officer in the society of physicists, when someone retires, and so on.① All of the little professional initiations and celebrations, like those in the family, are rituals. Rituals transform a graduate student into a research assistant, and give his fellow researchers a formal way to accept him as a co-worker. A ritual may be large, such as a university's graduation ceremony, or small, such as, in a research institute, a leader's introduction of a new colleague.②

Anthropology and medicine

Once we realize that culture shapes all of people's choices and actions, we can see that culture is a key to understanding medicine—how medicine varies in different places and changes in time. Medical anthropologists learned a long time ago that people in two cultures do not have exactly the same diseases. There are many ways to divide all physical abnormalities into diseases, and people in different cultures make different choices.

If we assume that the *experience* of disease is the same everywhere, we can make serious historical mistakes. For instance, it is common for people reading old books such as *Shanghan lun* (Treatise on cold damage disorders) to think

---

① Galison 1987 is a well-known study of these and related questions.
② See, among others, the writings of Bruno Latour.

that *re* means "fever." In biomedicine, *re* is a body temperature higher than normal. But in early Chinese writing, *re* is usually not a temperature on the surface of the body that the doctor measures, but hot feelings inside the body that the patient experiences and tells the doctor about. ① It is the opposite of *han*, "chills," cold feelings inside the body.

In the *Shanghan lun*, and even in traditional medical books of the 19th century, *shanghan* itself is usually not typhoid—the only meaning that modern dictionaries give—but a large group of different diseases that involve different kinds of fever. That is why I translate *shanghan* literally as "Cold Damage Disorders," to remind readers that the experience of disease in China before modern times was not the same as today. ②

People learn as they grow up what is a sickness and what is health. They also learn from their parents how to be sick-that is, how people are expected to behave when they feel sick, who to ask for help, and so on. The ways that doctors classify disease (nosology) changed greatly during imperial times. Since the introduction of Western medicine, nosology has changed even more quickly.

Today when we think of medicine we think of doctors. But anthropologists remind us that many different kinds of people treat sickness. The first step of health care is self-care. If you

---

① *Re* is used rather consistently in this meaning. *Fare* first appears in the *Shanghan lun*, but in most cases its usage is too vague to be sure how often at that stage it referred to hot feelings, and how often to high body temperature.

② I give many examples of such differences in Sivin 1987. For the changing meaning of a single Chinese disease name, see the detailed study in Smith 2008.

wake up in the morning with a headache, you don't go to the doctor, you might simply take an aspirin or *zhengtianwan*. If you are not sure what is wrong, you ask the advice of your mother, wife, friend, or someone else you trust. That is usually the second level of health care. If you still don't feel better, you may go to a drug store and ask someone what he recommends. Then, if those levels of health care fail, you may go to the clinic. Health care almost always includes many levels, with professional doctors only at the highest levels.

This was equally true in traditional China. Most people, in fact, never saw a physician. There were too many people in China, most people had too little money, and there were too few doctors. A farmer's wife in the country was likely to take her sick child to a neighbor who gathered herbs in the mountains, to a traveling curer, or to a popular priest. For a historian, understanding the value of their therapy is not easy—but understanding the value of a doctor's therapy is not easy either. Educated people today often reject any report of medical success that laboratory tests have not proved. But medical anthropology showed long ago that that is too narrow a standard to be useful.

One summer morning many years ago, when my wife and I were in Cambridge, England, I woke up, stood up, and immediately fell flat on my face. This seemed odd, but I stood up again, and fell flat on my face again. I could no longer control my body. I asked myself what was wrong. If it had been a headache, I would have recognized it and taken an aspirin. In this instance I had no idea what had happened. The only thought that occurred to me was "if this is permanent, it's going to be awfully inconvenient!" Before long, my wife noticed me lying on the floor. The next step was indeed asking her advice. As a result, she telephoned a friend, and quickly found out that the a strange new virus had also sickened several friends, and that if I waited 48 hours it would be gone. Since we were no longer worried, we did not need to call a physician. A day later, as the symptoms began to fade, we were confident that no therapy would be necessary. That was the end of that.

Let us imagine a similar case in ancient China. Someone who could not stand up in the morning would feel, as I did, that he could no longer control his body, that he was helpless, but it was not exactly the same feeling. He would have learned a different set of possible causes to consider than I did. A cause that people often thought about long ago—and still think about in parts of China—is that a spirit had taken over his body. His wife might invite a popular priest to drive the spirit out. Odd though it seems today, people believed that priests had the authority to control spirits. Whether this control was an objective fact or not does not matter; people believed it. That belief was often more powerful than a drug. A priest's ritual was designed to convince the patient, as my wife's telephone call convinced me, that his life was not really out of control, that he could expect to return to normal before long. Belief can indeed overcome real physical illness, because belief can cause real physical illness. The body responds to belief and meaning. That is the greatest discovery of medical anthropology.

Doctors, Chinese and Western, also use belief to cure

illness. The way a physician today acts, the language he uses, the way he uses instruments, all cause change in the patient. Instruments are not only tools of diagnosis and therapy. They are also symbols of the physician's knowledge and skill, symbols that he can use and other people cannot use. They are part of a ritual that includes the physician and the patient, what we might call a ritual of science. The ability to use science, which people respect, gives the doctor great authority to impose order on disorder. That is why anthropologists believe that to fully understand modern medicine we need to analyze both its technical and symbolic value.

In fact there are three kinds of things that cure illness.

First, most people who fall sick recover, even when they do not receive therapy. The body responds directly to abnormality, when it can, by becoming normal again. Doctors from Hippocrates (460-ca. 370) on have realized that their most powerful methods simply support the patient's own recovery. This is also an important theme in *Huangdi nei jing* and other Chinese classics. For most illnesses, all physicians need to do-or can do-is encourage recovery, and treat symptoms. We might call this tendency to recover the body's own response.

Second, both Chinese and Western medicine are based on the assumption that physical and chemical therapy can affect the body's processes and overcome disease. We might call this the technical response.

Third is the body's response to ritual and other meaningful symbols. Some anthropologists call this the meaning response. All three of these responses may be taking place at the same time.

Not many anthropologists study the past, but these and other insights have turned out to be equally valuable for historical research. To sum up, an adequate analysis of therapy includes the body's ability to heal itself, the effect of the ritual and symbolic circumstances of therapy, and the value of the technical methods.① A doctor who is trained to understand all three components of therapy is likely to be a more effective therapist than one who understands only the technical part. The same understanding is useful for historians of medicine studying the past. In the history of astronomy, similarly, knowing how people at the time thought about the sky and events in it is as important as understanding techniques of prediction. For instance, Sun Xiaochun and a colleague wrote a very valuable book about how the sky looked to people in the Han dynasty.

Sociology

The basic idea of sociology is that in order to understand the activity of individuals, we must comprehend the structures, norms, and values of the society in which they live. Every society has its own understanding of what the normal relationships between people are, of how people should behave, and of what they should consider good and bad.

---

① The best discussion of these three components of medical efficacy is Moerman 2002.

Individuals do not behave exactly the way other people expect, and people do not expect exactly the same thing. Those structures, norms, and values are ideals, important in teaching people how to live. Parents teach them to their children, teachers teach them to their students, and even television teaches them to people who watch it. That is why we recognize that Chinese behave in certain ways, and Americans behave in other ways. We can see that Chinese-Americans behave in ways that reflect both what their Chinese parents teach them and what they learn from other non-Chinese Americans. I will discuss a couple of sociological topics today, namely the concept of profession and changes in nosology.

The Chinese words for "occupation" and "profession" are both *zhiye*, but in Western thought about society their meanings are quite different. That is because of cultural differences over the centuries. According to sociologists, any *zhiye* is an occupation, but only a *zhiye* that requires a special high level of education, and results in a high social status, is a profession.

In the early European universities (from about 1200 on), there were only three groups who had more than a basic education: priests, lawyers, and doctors. Unlike other occupations, these groups themselves gave people in them three special powers: the groups themselves controlled who entered their profession, they determined how much money members were paid, and they could prevent unqualified people from practicing their profession. Since the 19th century, priests lost much of their social authority. Today doctors and lawyers are the two model professions. Other groups tried to gain the same high status, such as scientists. They used the same methods—postgraduate education, and so on. But most scientists are employees of universities and other institutions. They do not set their own incomes, and cannot control who does science.

Sociology of Medicine

In most European countries and the United States, doctors and lawyers still have the highest prestige of all occupations. Still, the peculiar forms of public or private medical insurance that are widespread in those countries have taken away physicians' control of their incomes, and made average incomes much smaller. More and more lawyers are employees rather than working as partners with other lawyers. So the autonomy and professional character of both occupations is disappearing. In some European countries it disappeared some time ago, especially in Socialist countries where such people are government employees.

Because of the special character of Chinese society, medicine is an occupation, not a profession in this historical sense. In ancient times, anyone could practice medicine, regardless of education, and most doctors were trained as disciples of older doctors rather than in special schools. They did not need a special degree or license to practice. For many centuries the government gave medical examinations, but they were usually only for positions in the palace medical service, not for public practice. The members of the palace medical service were indeed officials, but their positions were not high. ①

Doctors who were not officials often complained about the competition of what they called quacks, but they could not prevent them from working. The income of a doctor depended on his personal reputation, and to some extent on the reputation of his teacher. But the reputation of a doctor did not necessarily depend on therapy. Many experts on medicine were famous because they were high-ranking non-medical officials or exceptional writers. Such people mainly treated their friends, and did not try to make a living from medicine. Two well-known examples, from the Song dynasty, are Shen Kuo (or Gua, 1031 - 1095) and Su Shi (or Dongpo, 1037 - 1101). But those who earned a living as physicians occupied many different positions in society, and had to compete with religious and popular healers of all kinds. Before 1920, doctors were not organized as a single occupation.

In China since Liberation, it has been the government that determines who is a doctor, that sets his income, and that decides who is no longer qualified to practice. Many doctors who work in private practice full time or part time are able to increase their incomes, but the government still regulates most aspects of medical practice.

To sum up, the careers of doctors varied all over the world. Their social standing differed from one country to another, and

---

① See the details in Liang Chün 1995.

changed in each society. My point about professions and occupations is that sociology can aid in understanding the differences in the authority of doctors in different societies. Authority, we have already seen, is an important factor in their ability to cure illness.

Sociology of Disease

Another important issue in sociology is how diseases are defined. Long ago, anthropologists discovered that different cultures had different diseases. It is not true that scientific research alone determines what is a disease. Even in neighboring European countries, doctors' understandings of illness and therapy differ considerably.① In any one country, the understanding of what is a disease changes continuously and often quickly. Let me give a couple of examples from the United States; in such cases the situation in Europe is different.

• In the 18th century, when Christian religion had great power in the United States, drinking too much alcohol was a sin, condemned by priests. In the 19th and early 20th centuries it became illegal, a crime punished by judges. When I was a child, people who were drunk in public were often arrested. In the late 20th century, it became a disease and its

---

① Payer 1988 has ingeniously studied this question for England, France, the United States, and Germany. Despite the great changes that have taken place in all four countries in the last twenty years, no one has done a similar study since.

name was changed to "alcoholism."

This was not because doctors discovered a cure. Even today they cannot explain the cause of alcoholism, and they are unable to cure it. Sociologists argue that the transformation of drunkenness from a sin to a crime was the result of shrinking religious authority and growing state authority. Perhaps the shift from drunkenness to alcoholism happened because punishing people for drinking too much was not effective. "Drunkenness" suggests that someone is a bad person; "alcoholism" means that it is a sickness, a problem that is not the fault of the individual. The inability of medicine to cure it does not matter. Medicine cannot cure many other serious physical problems, such as cancer. Many people hope that, though it cannot cure alcoholism, some day medicine will succeed where the law failed.

- In the United States, senility or senile dementia used to be quite frequent. In this disease, old people lose their memories, become very confused, and often do not even recognize members of their own families. At the same time, Alzheimer's disease was a very rare disease in which young people have the same problem. The symptoms were almost identical; the main difference is that young people very seldom develop them. Neither disease is curable. Remarkably, over the past twenty years, doctors have stopped diagnosing senility. They now use "Alzheimer's disease" even when their patients are very old. It has changed from a rare diagnosis to a common one, and doctors no longer decide that the disease is senility. The change did not come from new scientific research.

Rather, it seems to be a shift from a word that people considered bad to one that seems neutral and scientific.

It is easy to find many examples of these kinds.[①] They suggest that new diseases, and changes in the names of diseases, are not only based on improvements in research. Physicians, like all other members of society, respond to values that are widespread in their society. If they do not, they cannot be very successful in curing sick people.

If a doctor believes that drinking too much is bad behavior, and not a disease, he can play no part in the treatment of alcoholism. If he insists on calling a patient senile when other doctors talk about Alzheimer's disease, it is he and not they who will seem to be wrong. Examples like these indicate that naming diseases is a social phenomenon in which doctors and laymen influence each other until they agree. In fact, a certain kind of diagnosis sometimes becomes a fashion. That is one of many ways in which sociology is necessary for fully understanding medicine.

Conclusion

Anthropologists and sociologists had different origins and developed different concepts. But over the past hundred years, as they read each other's work, they became aware of many

---

① See Aronowitz 1998 for many such analyses of changes in diseases. See also Kleinman 1986.

similarities. Eventually some anthropologists began studying cultures as sophisticated as their own—for instance, the culture of China. One of my teachers used to say that Americans could learn a good deal if Chinese anthropologists studied them. Many anthropologists now study their own cultures (the first for China was Hsu 1952).

At the same time, sociologists found that their techniques were useful in studying societies different from their own. It is not surprising that both groups eventually learned to use both the anthropological concept of culture and the sociological concept of society. Of course, institutions do not change quickly. The two still have different professional associations, journals, and so on.

In my next lecture, on popular culture, I will take up other uses for anthropology and sociology in the history of science and medicine.

References

Aronowitz, Robert A. 1998. *Making Sense of Illness. Science, Society, and Disease*. Cambridge History of Medicine. Cambridge University Press.

Galison, Peter. 1987. *How Experiments End*. Chicago: University of Chicago Press.

Hsu, Francis L. K. (Xu Langguang). 1952. *Religion, Science and Human Crises: A Study of China in Transition and its Implications for the West*. International Library of Sociology and Social Reconstruction. London: Routledge & K. Paul.

Kleinman, Arthur. 1986. *Social Origins of Distress and Disease. Depression, Neurasthenia, and Pain in Modern China*. New Haven: Yale University Press.

Liang Jun. 1995. *Zhongguo gudai yizheng shilue* (Outline history of ancient Chinese medical administration). Huhhot: Neimenggu Renmin Chubanshe.

Latour, Bruno. 1987. *Science in Action: How to Follow Scientists and Engineers through Society*. Milton Keynes: Open University Press.

Latour, Bruno, & Steve Woolgar. 1986. *Laboratory Life: The Construction of Scientific Facts*. Princeton University Press.

Moerman, Daniel. 2002. *Meaning, Medicine and the "Placebo Effect."* Cambridge Studies in Medicine Anthropology. Cambridge University Press.

Payer, Lynn. 1988. *Medicine & Culture: Varieties of Treatment in the United States, England, West Germany, and France*. New York: Henry Holt.

Sivin, Nathan. 1987. *Traditional Medicine in Contemporary China. A Partial Translation of Revised Outline of Chinese Medicine (1972) with an Introductory Study on Change in Present-and Early Medicine*. Science, Medicine and Technology in East Asia, 2. Ann Arbor: University of Michigan, Center for Chinese Studies.

Smith, Hilary. 2008. "Foot Qi: History of a Chinese Medical Disorder." Ph. D. dissertation, History and Sociology of Science, University of Pennsylvania.

Sun Xiaochun; Jacob Kistemaker. 1997. *The Chinese Sky during the Han. Constellating Stars & Society*. Leiden: E. J. Brill.

Using Studies of Popular Culture

N. Sivin
2009. 4. 15

Popular culture was the basis of health care, agriculture, the making of ceramics, the manufacture of metal tools, and many other crafts. By using critically religious writings, literature, historical records, and collections of jottings, we can put together a picture of science, technology, or medicine in popular culture. In ancient China, the physician's learning made him an expert in the dynamics of the cosmos, and the popular priest's knowledge of spirits gave him influence in the bureaucracy of the gods. *Zhu bing yuan hou lun*, discussing a disease it calls "*chan hou xue yun men*" (dizziness and nausea after delivery of a child), and *Qianjin yi fang* recording the therapy for "*chan yun*" (dizziness after childbirth), show that prohibition, taboo and ritual were too important to disregard. Anthropology and sociology offer the tested and proven tools for inquiry into popular culture.

3 Using Studies of Popular Culture

Most of our knowledge of science and medicine in China is about the work of the elite, the ruling class of China.① Mencius, before the beginning of the Chinese empire, described the difference between those who governed and the rest of the population:②

> Some labor with their hearts and minds; some labor with their physical strength. Those who labor with their hearts and minds govern others; those who labor with their physical strength are governed by others. Those who are governed by others feed others. Those who govern others are fed by others. This is a principle that everyone in the realm accepts.

As you know, this gap between rulers and ruled continued throughout the history of the Chinese empire. There were many fundamental differences. The whole class of gentlemen (*shi*) were eligible to become officials, and had many special privileges. Until at least the Song dynasty, it was mostly people of this class who owned large amounts of land and collected rents. The most important difference between the elite and commoners was that the elite were educated; they could read and write. If we study history, we study it from

---

① This lecture is respectfully dedicated to the memory of Professor Chen Meidong.
② *Mengzi*, 3A.4：或劳心，或劳力。劳心者治人，劳力者治于人。治于人者食人，治人者食于人，天下之通义也。

books and documents that they wrote. That means that to a large extent we see the world around them through their eyes. For instance, when Mencius talks about "a principle that everyone in the realm accepts," he did not mean that poor farmers, who worked their whole lives to feed others even when their own families starved, accepted this principle. There is no evidence that Mencius asked poor farmers about principles.

From about the Northern Song period on, a growing number of people who were not *shi*, mainly merchants 商 and artisans, could read and write to some extent, and from the Ming period on, a few such people were as well educated as officials. But most commoners, especially poor farmers, were still illiterate.

What we know about the uneducated comes mostly from the writings of those who governed them, the people they fed. As historians, we realize that these writings are not objective. We have to identify the bias of each writer. But if we do that, we can put together from many writings a general picture of life among people who did not belong to the elite.

I will talk today mainly about health care, since I have been working for many years to put together a general picture of it. Nevertheless, the methods I have used are also useful for many other technical topics.

Practitioners of popular therapy, as I mentioned in my last lecture, ranged from doctors less educated than those who practiced classical medicine,① to farmers who gathered herbs in the mountains, traveling curers, teachers of meditation to Daoist priests, each with different ideas of health and sickness, and with different kinds of therapy.

We find in medical books a great many discussions of the illnesses of ordinary people. There is a great deal of information about the therapy of healers who were not physicians. Because it played such a large part in the health care of the Chinese population, I suggest that it is worth while to examine it closely.

In addition to medical books, there are many other sources of information. Because curing illness is important in every religion, I have found much material in religious writings, especially the Daoist scriptures and, to some extent, the Buddhist scriptures. But there are also useful data scattered in every kind of literature, including history books and collections of jottings. ②

By using all of this information critically, we can put together a fairly clear picture of medicine in popular culture. Since it differs greatly from elite medicine, I will describe the basic ideas of elite medicine first, and then compare the ideas of popular medicine.

---

① I use "classical medicine" to mean Chinese medicine before the 20th century. Because influence from foreign medicine was small, it is quite different from the modern traditional Chinese medicine (TCM).

② Among the useful reference books for this work are Chen Bangxian 1982, Tao Yufeng et al. 1988, and Zhongyi Yanjiuyuan 1989.

The world view of elite medicine

In the two and a half centuries after 250 B. C. , a number of important philosophers produced a world view that included nature, politics, and the human body.① The world of nature, the universe, operated according to regular cycles. Philosophers described them using the technical language of yin-yang and the Five Phases. The daily cycle of light and darkness, the yearly cycle of warmth and cold, the movements of the sun, moon, planets, and stars were regular and eternal. The activities of a well-governed state followed the same rhythms as nature did, and therefore in theory it could also continue forever. The emperor, as "the son of heaven," was responsible for maintaining the relationship between the natural order and the political order—that is what good government meant.

The body, in the same way, was healthy as long as it remained in harmony with nature's cycles of change. That required living in a correct way, avoiding excess in behavior, thought, or emotion. Illness struck people who did not maintain such a balance. The work of the physician was to determine what the imbalance was, and how to restore

---

① The most important of these were *Lü shi chunqiu* (ca. 239), *Huainanzi* (ca. 139), and *Chunqiu fan lu* (ca. 156 and later). For a synthesis that includes nature, politics, and medicine, see *Huangdi neijing* (first century B. C. ?). There is a detailed analysis of the synthesis and its formation in Lloyd & Sivin 2004, appendix.

balance. His main tool of technical analysis was the philosopher's concepts of yin-yang and the Five Phases, and other concepts such as the Six Warps based on them. ① From his teacher or from books on material medica and other kinds of therapy he learned what resources he could use to correct the imbalances he found.

The skill of the elite physician, in other words, was in analyzing the processes of the universe and of the body, and determining how to bring them into the state of harmony that we call health. ②

Popular Medicine

Popular medicine includes all the kinds of therapy that were part of popular culture. That term needs a definition. It is not the opposite of elite culture, but a larger category. It is more accurately to translate it *pubiande wenhua* rather than *minjian wenhua*. ③ It is a view of the individual, society, and

---

① Every medical-school textbook of "basic doctrines" explains these and other concepts and their use. On differences between the ancient and present-day meanings of concepts, see Sivin 1987, 43 – 196.

② There was no single word in ancient Chinese for health. Medical authors used several words such as *pingren* to describe healthy people. *Jiankang*, the word common today, was borrowed in modern times from Japanese. On health and disorder in early medicine, see Sivin 1987, 95 – 111.

③ Today "popular culture" in consumer societies mainly refers to preferences in spending money. In English "popular" was used from the 15th century on to refer to "vulgar" or low-class culture, but from late 18th century the main meaning of "popular" in scholarship has been "widespread."

nature that everyone in Chinese society before modern times learned as they grew up。 Everyone was familiar with it, whether they were educated or uneducated, powerful or powerless. Male members of the elite, as part of their education, learned elite culture as well. In some circumstances—for instance, as officials—they showed contempt for popular culture. At other times—for instance, when they lived in their native villages—they actively supported it. "Popular medicine" is therefore a more accurate term than "folk medicine." Today I will talk about the popular medicine of the Han people. China's ethnic minorities generally had their own beliefs and practices.

Popular medicine, then, means an understanding of the body, health, illness, and therapy that Han people shared. Rather than being based on elite medicine's view of cosmic processes, it was based on understanding the society of spirits. Analyzing this set of ideas makes it possible to explain how popular medicine worked.

Ordinary people believed that alongside human society there lived a society of spirits, which we can divide into gods, ghosts, and ancestors. When someone died, it was the family's responsibility to care for that person's spirit with certain rituals and offerings. Doing that properly made the dead person an ancestor. Such a spirit would be benevolent toward his descendants. A ghost was a spirit that was not properly cared for, and therefore might harm its descendants and other people. A god was a spirit that is benevolent not only toward its own descendants but toward a group, a community or a whole population.① Curing illness is one important kind of benevolence. When this happened to people of more than one family, who then prayed or made offerings to that spirit, then people might begin to think of it as a god. Such a spirit, if its worship became widespread, would be generally acknowledged as a god, and, for that reason, appointed to an official post in the celestial civil service. That bureaucracy was organized in the same way as the imperial civil service, and had similar authority over the spirit world.

People of different families often did not agree about who became a ghost or even a god. My ancestor might be your ghost or our village temple's god. Like imperial civil servants, these are not different kinds of beings, but different functions of dead people.

Original sources that discuss popular medicine reveal that most disease was possession. That is, dangerous spirits or ghosts invaded the body and interfered with its normal processes. This could happen for various reasons. The most usual reason was immorality or bad behavior. Neglecting the spirit of a dead member of your family is one example. The result was pollution. A healthy person was spiritually and morally clean. If he became unclean, spirits could invade his body and make him sick.

Popular healers—for instance, priests—were experts in the bureaucracy of the gods. They could find out which spirit

---

① Not all gods were the spirits of dead people; some personified aspects of the universe.

was causing trouble, and knew how to end the possession. To understand this, think about the local government office in imperial China. Most ordinary people knew very little about it, and had no contact with it. But sooner or later, a messenger might come from the *yamen* on some unimportant business, and would want some money. If you didn't have any money, what could you do? If you went to the *yamen* and accused him, you would probably get into trouble. But if you knew someone in your village who had friends in the *yamen*, you could ask that person to help you. That was like the skill of the popular priest: he knew something about the celestial bureaucracy, and how to use rituals to find help. Once he found out which spirit had entered your little brother's body, your family would take part in the priest's rituals to get rid of the pollution and drive the spirit out.

That is what almost everyone in China believed before modern times. The work of anthropologists all over the world has shown that belief can cause illness, and belief can cure it. Applying their insights opens the way to understanding fully the history of health care in China.

Now let me sum up my comparison of elite and popular medicine. First, they are based on different understandings of the body's relations to what surrounds it. Elite physicians understand health as the body's harmony with the dynamic rhythms of the universe, illness as a break in that harmony, diagnosis as determining what the disharmony is, and therapy as various ways of re-establishing harmony. Popular priests understand health as the body's moral and spiritual purity, illness as possession by a spirit due to pollution, diagnosis as identifying the possessing spirit, and therapy as various rituals to restore the body's purity. This purity in turn restores the harmony of the individual with the people around him. To put it another way, the physician's learning makes him an expert in the dynamics of the cosmos, and the popular priest's knowledge of spirits gives him influence in the bureaucracy of the gods.

Efficacy

The next obvious question is how medical anthropology and the sociology of medicine help us to understand therapy. What have they learned about the effectiveness of popular medicine? I suggested in the last lecture that three kinds of things combine to cure illness:

- the ability of the body to recover spontaneously, without therapy: the body's own response.
- the ability of drugs and other therapy to bring about chemical and physical change in the body: the technical response.
- the body's response to ritual and other meaningful symbols, which can also bring about chemical and physical change in the body. Some anthropologists call this the meaning response.

The first two are not difficult to understand, but it may help to give an example of the third. Several psychiatrists have studied Chinese popular therapy and, because of their psychological specialty, concluded that ritual can cause psychological change in patients.① They argue that ritual is a kind of primitive psychotherapy. That is true to some extent, but it is too narrow an explanation. Therapy affects not only the mind and the body, but—because in each culture people learn how to be sick—it also affects social relations. My detailed example will make this point clear.

A very important early book on the causes of illness, *Zhu bing yuan hou lun* (Origins and Symptoms of Medical Disorders, A. D. 610), discusses a disease it calls *chan hou xue yun men* (dizziness and nausea after delivery of a child). It is a serious disease, which can kill the mother. This book, written by an imperial physician, describes and explains the illness in the language of classical medicine, emphasizing the balance of yin and yang vitalities (*xueqi*) in the circulatory system:②

> The manifestations of dizziness and nausea are worry and inner tension so bad that one's *ch'i* seems about to expire.

---

① The best-known are Kleinman 1974, 1980, and 1986. The first contains papers by several psychiatrists.

② *Zhu bing yuan hou lun*, 43: 230b: 运闷之状, 心烦气欲绝是也。亦有去血过多, 亦有下血极少, 皆令运。若产去血过多, 血虚气极, 如此而运闷者。但烦闷而已。下血过少而气逆者, 则血随气, 上掩于心, 亦令运闷。则烦闷而心满急。二者为异。
亦当候其产妇血下多少。则知其产后应运与不运也。然烦闷不止, 则毙人。
然产时当向坐卧, 若触犯禁忌, 多令运闷。故血下或多或少。是以产处及坐卧须顺四时方面, 避五行禁忌。设有触犯, 多招灾祸也。

Xueqi: It is a common mistake to think that *xue* 血 always means physical blood. Medical authors divided the *qi* in the body into two aspects, one yin, which they called *xue*, and one yang, also called *qi*. See Sivin 1987, 50 - 60.

There are also cases in which too much bleeding or very scanty bleeding may bring on dizziness. . . .

The text describes in detail different types of the disease and how to distinguish them. It goes on to point out how dangerous it can be:

If the feelings of upset and tension do not stop, they may kill the patient.

The book then goes on to explain what causes these difficulties:

If a woman violates the prohibitions concerning the direction she should face when sitting or lying down during labor, dizziness will usually result. The downflow of blood will be either excessive or deficient. When her direction violates a prohibition, this will usually bring on dizziness and nausea, and bleeding may be overly copious or scanty. On account of this, where childbirth takes place, and whether she lies down or sits up, must be appropriate to the season and the direction, to avoid violating the prohibitions according to the Five Phases. Violations will generally bring disaster.

This discussion says clearly that the ultimate cause of this disease is bad behavior, namely breaking certain rules about the four seasons and the Five Phases. It identifies these rules as a prohibition or taboo. Prohibitions of this kind do not appear often in classical medical books. They indicate that this disease comes from the world of popular medicine, where facing certain directions for something like childbirth is important.

This is not mere nonsense. As any modern physician

knows, dizziness, nausea, and changes in blood flow are common symptoms of stress and anxiety. Let us say that long ago a midwife was helping a woman whose labor was difficult. The midwife might come to believe that the labor was difficult because the woman was facing in a forbidden direction. If she warned her that she might have a dangerous disease, the result naturally could be great anxiety and the other symptoms of this disorder.

Now let us examine therapy for this disorder from an anthropological point of view. In the famous *Bencao gangmu* (Systematic materia medica, 1596), a formula from an earlier source uses blood from childbirth to treat a disorder called "abnormal blood loss and dizziness after the birth of a child" (*chanru xue yun*): "Take concentrated vinegar and mix it with blood collected at the time of delivery [and coagulated], the size of a jujube 枣, and give it to the patient to eat."①

Again the use of human blood suggests that this formula was taken from popular medicine, but it does not answer the next question an anthropologist will ask: what was the ritual? Today people simply stop for a moment, swallow a pill, and go on with what they were doing. But a popular therapist used ritual, a pattern of action and speech that embodied powerful symbolic meanings. We do not know what symbols and actions were used in this case to calm the patient's anxiety and stop her dizziness.

---

① *Bencao gangmu*, 52: 100, 取酽醋, 和产妇血, 如枣大, 服之。This source credits the formula to *Taiping sheng hui fang* 太平圣惠方, but it does not appear in the 1958 Renmin Weisheng Chubanshe edition, pp. 2493 – 97。

Discussion in Cooper & Sivin 1973, 238 – 239.

We do have a ritual from another source. Sun Simiao's *Qianjin yi fang*, of the late 7th century, contains a remarkable collection of methods that Daoist priests① used to drive out spirits that were possessing people and causing disease. This "Manual of Interdiction (*Jin jing*)" includes the details of a ritual for curing "dizziness after childbirth (*ch'an yun*)." From it we can learn something about how such rituals can be effective, even if they do not seem scientific: ②

> Take seven cloves of garlic. On the first day of the first month, facing due east, have the wife recite [the following interdiction] once; then the husband is also to chant it once. One at a time, the husband swallows a clove of garlic and seven sesame seeds. Then he walks due east and repeats the formula a full seven times. [The husband and wife] may not see anything polluting or loathsome. In learning and practicing [this interdiction] avoid seeing corpses or mourners, or it will not be effective:
>
> I tread the mainstays of the sky, roam the Nine Realms.
> Hearing of your difficulty in childbirth, I come in search,
> To dismember and slaughter unlucky [spirits],

---

① One method contains the words *Wu wei tianshi jijiu* 吾为天师祭酒, "I am a libationer—an ordained priest—of the Heavenly Masters"; *Qianjin yi fang*, 29: 347b.

② *Ch'ien chin i fang*, 29: 22b‐23a: 取蒜七瓣。正月一日，正面向东。令妇人念之一遍，夫亦诵一遍。次第丈夫吞蒜一瓣，吞麻子七枚，便止。丈夫正面向东行，诵满七遍。不得见秽恶。守持之法，不用见尸丧，见即无验。

吾蹋天刚，游九州。闻汝产难，故来求。斩杀不祥，众喜投。母子长生，相见面，不得久停留。急急如律令。

The last sentence is a conventional way that a Daoist priest voices his authority. On the celestial mainstays, the cosmic meridians that bind the stars together, see Schafer 1977, ch. 12.

Replace them by every felicity.
To mother and child, long life in mutual regard.
Impermissible [for spirits] to tarry long.
Quickly, quickly, by lawful order.

The last two lines send the spirits on their way, exerting the Daoist priest's authority in the celestial bureaucracy.

Here we can see the main point of symbolic curing. Before the ritual, the woman is an isolated and helpless victim of possession by a dangerous spirit. The ritual transforms her into a person protected by a priest who has the authority to drive out or even kill menacing ghosts.[①]

But how does this ritual prevent dizziness after childbirth? What can it change? In order to answer that question, we have to look at not only the wife's mind, but her relationship with her husband. Why, in fact, does the husband play a larger role in this ritual than his wife?

Keep in mind that, in the 7th century, in fact until recent decades, when a Han woman was married, she ordinarily moved into the house of her husband's parents. Her main responsibility was serving them, at least until she had a male child. Traditional society did not encourage husbands to think about their pregnant wives' anxieties. If a husband showed too much affection for his wife, he might offend his parents.

This ritual focuses the husband's attention on the dangers that his wife will face in childbirth. The chanted interdiction lets him openly express his concern. If he does this sincerely, a

---

① See the classic study of a similar case by Claude Levi-Strauss (1949).

References

Chen Bangxian. 1982. *Ershiliushi yixue shiliao huibian* (Collection of materials for the history of medicine from the twenty-six Standard Histories). Beijing: Zhongyi Yanjiuyuan, Zhongguo yishi wenxian yanjiusuo. Completed 1964.

Cho, Philip S. 2005. Ritual and the Occult in Chinese Medicine and Religious Healing: The Development of *Zhuyou* Exorcism. Ph. D. dissertation, History and Sociology of Science, University of Pennsylvania.

Cooper, William C., and Nathan Sivin. 1973. "Man as a Medicine: Pharmacological and Ritual Aspects of Traditional Therapy Using Drugs Derived from the Human Body." In Nakayama and Sivin 1973, 203 – 272.

Kleinman, Arthur, editor. 1975 (publ. 1976). *Medicine in Chinese Cultures. Comparative Studies of Health Care in Chinese and Other Societies: Papers and Discussions from a Conference Held in Seattle, Washington, U. S. A., February* 1974. DHEW publication (NIH) 75 - 653. U. S. Dept. of Health, Education, and Welfare, Public Health Service, National Institutes of Health.

Kleinman, Arthur. 1980. *Patients and Healers in the Context of Culture: An Exploration of the Borderland between Anthropology, Medicine, and Psychiatry*. Comparative Studies of Health Systems and Medical Care, 3. University of California Press.

Kleinman, Arthur. 1986. *Social Origins of Distress and Disease. Depression, Neur-asthenia, and Pain in Modern China*. Yale University Press.

Kvale, Steinar, & Svend Brinkmann. 2009. *InterViews: Learning the Craft of Qualitative Research Interviewing*. 2d edition. Los Angeles: Sage Publications.

Levi-Strauss, Claude. 1949. "L'efficacite symbolique." *Revue de l'histoire des religions* 135: 5 – 27. Translated as "The Effectiveness of Symbols" in *Structural Anthropology* (New York, 1963), pp. 186 – 205.

Lloyd, G. E. R., and Nathan Sivin. 2002. *The Way and the Word. Science and Medicine in Early China and Greece*. New Haven: Yale University Press.

Makagon, Daniel, & Mark Neumann. 2009. *Recording Culture: Audio Documentary and the Ethnographic Experience*. Los Angeles: Sage.

Nakayama, Shigeru, and Nathan Sivin, editors. 1973. *Chinese Science. Explorations of an Ancient Tradition*. MIT East Asian Science Series, vol. 2. Cambridge, Mass.

Schafer, Edward H. 1977. *Pacing the Void. T'ang Approaches to the Stars*. University of California Press.

Sivin, Nathan. 1987. *Traditional Medicine in Contemporary China. A Partial Translation of Revised Outline of Chinese Medicine (1972) with an Introductory Study on Change in Present-day and Early Medicine*. Science, Medicine and Technology in East Asia, 2. Ann Arbor: University of Michigan, Center for Chinese Studies.

Spradley, James P. 1979. *The Ethnographic Interview*. New York: Holt, Rinehart and Winston.

Tao Yufeng; Zhu Bangxian; Hong Pimo. 1988. *Lidai biji yi shi pie lu* (Classified anthology of medical matters in the collected jottings of various periods). Tianjin: Tianjin Kexue Jishu Chubanshe.

Zhongyi Yanjiuyuan, Zhongguo yishi wenxian yanjiusuo. 1989. *Yixueshi wenxian lunwen ziliao suoyin* 1979 – 1986. (Index to essays and materials on the history of medicine and medical literature. Collection 2, 1979 – 1986). Beijing: Zhongguo shudian.

北京大学出版社教育出版中心
部分重点图书

一、大学教师通识教育系列读本（教学之道丛书）
　　给大学新教员的建议
　　规则与潜规则：学术界的生存智慧
　　如何成为卓越的大学教师
　　教师的道与德
　　给研究生导师的建议
　　理解教与学：高校教学策略
　　高校教师应该知道的120个教学问题

二、大学之道丛书
　　后现代大学来临？
　　知识社会中的大学
　　哈佛规则：捍卫大学之魂
　　美国大学之魂
　　大学理念重审：与纽曼对话
　　一流大学卓越校长：麻省理工学院与研究型大学的作用
　　学术部落及其领地：知识探索与学科文化
　　大学校长遴选：理念与实务
　　转变中的大学：传统、议题与前景
　　什么是世界一流大学？
　　德国古典大学观及其对中国大学的影响
　　学术资本主义：政治、政策和创业型大学
　　高等教育公司：营利性大学的兴起
　　美国公立大学的未来
　　公司文化中的大学
　　21世纪的大学
　　我的科大十年（增订版）
　　东西象牙塔
　　大学的逻辑（增订版）

三、大学之忧丛书
　　高等教育市场化的底线
　　大学之用（第五版）
　　废墟中的大学

四、管理之道丛书
　　世界一流大学的管理之道
　　美国大学的治理
　　成功大学的管理之道

五、学术道德与学术规范系列读本（学习之道丛书）
科技论文写作快速入门
给研究生的学术建议
如何撰写与发表社会科学论文：国际刊物指南
学术道德学生读本
做好社会研究的10个关键
阅读、写作与推理：学生指导手册
如何为学术刊物撰稿：写作技能与规范（英文影印版）
如何撰写和发表科技论文（英文影印版）
社会科学研究的基本规则
如何查找文献
如何写好科研项目申请书

六、北京大学研究生学术规范与创新能力建设丛书
学位论文写作与学术规范
传播学定性研究方法
法律实证研究方法
高等教育研究：进展与方法
教育研究方法：实用指南（第五版）
社会研究：问题、方法与过程（第三版）

七、古典教育与通识教育丛书
苏格拉底之道
哈佛通识教育红皮书
全球化时代的大学通识教育
美国大学的通识教育：美国心灵的攀登

八、高等教育与全球化丛书
高等教育变革的国际趋势
高等教育全球化：理论与政策
发展中国家的高等教育：环境变迁与大学的回应

九、北大开放教育文丛
教育究竟是什么？100位思想家论教育
教育：让人成为人——西方大思想家论人文与科学

十、其他好书
科研道德：倡导负责行为
透视美国教育：21位旅美留美博士的体验与思考
大学情感教育读本
大学与学术
大学何为
国立西南联合大学校史（修订版）
建设应用型大学之路
中国大学教育发展史

A great many historians of science use comparison as a tool for learning.① The possibilities of human experience are so wide that any civilization can use only a part of them. A scholar who grows up in France and studies only French history will probably have no idea what possibilities were realized in Asia and Africa. That is unfortunate, since those possibilities may offer her clues to understanding French culture that otherwise she would never know.

Reading Chinese journals makes it clear that, in this country as in others, comparison is often useful to historians. Today I will talk about several problems of comparison that it is valuable to think about.

Prerequisites

Many comparative publications have no influence at all, for reasons worth considering. For those of you who are thinking about how you might use comparison in your work, I need to remind you of some essentials. First, although we usually compare something in two different cultures, that is not the only way. A good deal of historical work is actually a matter of

---

① This lecture is respectfully dedicated to the memory of Professor Li Di.

comparing some idea or custom in the same culture at different times or different places. I am sure everyone here knows that the relative freedom of elite Chinese women in the Tang period was largely lost by the Ming and Qing periods. In those times, most elite women were secluded, uneducated, and half-crippled by foot-binding. To take another example, if we compare medical practice before and after the Northern Song period, we can see that great changes in almost every aspect of it took place in roughly the 11th century.[①]

Still, when we think of comparison we usually think of two cultures. To do studies of that kind, it is essential that we be familiar with both. Too often, it is narrow specialists in one culture who do comparative investigations of two. That often leads to unbalanced conclusions, for a simple reason. A culture that you know well always seems to be richer, fuller of fascinating ideas and activities, than one you only know a little about. If you have spent years studying Islamic culture, and have read only a few books on India, it is unlikely that you will recognize what the crucial similarities and differences are. Even if the books you have read are excellent, it is not likely that their authors were asking the questions that you want to answer. To make a balanced evaluation, you need to be equally well informed about both cultures. One way to do that is collaboration. Still, for a fruitful collaboration, both people need to learn a good bit about each other's field.

To know any society well, you need to know its language,

[①] On changes in medicine and other fields, see Sun & Zeng 2007.

or languages. Anyone who hopes that a high-quality journal or book publisher will accept his work has to use original sources and to know the work of all his predecessors. Graduate students at my university, if they want to study ancient Chinese mathematics, need to know not only classical and modern Chinese, but French and Japanese, because there are important publications in all these languages. Historians of medicine also need German, for the same reason.

On the other hand, if you are interested in studying Galileo (Galilue, 1564 - 1642), you have to be familiar with the forms of Latin and Italian that Galileo wrote, which are quite different from classical Latin and modern Italian. You will also need modern Italian, French, German, and English, because there are many essential studies in those languages. Those prerequisites are difficult. In fact, people who do not begin any language they need before graduate school may find it is too late to learn so many languages. But nothing is worse than to spend years working on a dissertation, revising it to make a book, and then finding that it cannot be published. On the other hand, once you have made this effort, your work can be among the world's best.

Examples of Comparison

I will give several examples of comparison to illustrate how widely it can be used. Some of my examples will show how wide study makes it possible to explain the differences you find. Let me begin with using a dictionary to make comparisons of meaning.

Not every student learns how to use what is called "a dictionary on historical principles." Such a reference book, for every sense of a word, quotes the earliest known example of its use. Their editors could not do an original investigation of every word, but if scholars use them critically they are extremely helpful. The standard dictionary of that kind for English is the *Oxford English Dictionary*, which most scholars simply call the OED. It is available online from many university libraries. There have been a number of such dictionaries in Chinese, including the excellent *Hanyu dacidian*.

Nature

Using books of this kind lets you discover important problems of comparison. One example is *ziran*. Modern Chinese-English dictionaries usually define it as the noun 名词 "nature"—because people use it today as equivalent to *daziran*—or as the adjective "natural." If we look up *ziran* in the *Hanyu dacidian*, we find only four ancient meanings: (1)

天然，非人为地，"natural"（from the *Laozi*）;①（2）不勉强，不拘束，不呆板，"unforced"（*Houhanshu*）;（3）不经人力干预而自由发展，"spontaneous"（*Yeshi*）; and（4）犹当然，"naturally（*Beishi*）." According to this dictionary, in classical times the meanings of *ziran* did not include "nature" as a noun.

Why not? Because, apparently, at that time Chinese did not need a word for the physical world or cosmos. Why should we assume they needed it? In fact, a recent study of the origins of modern Chinese words revealed that *ziran* was first used for "nature" in 1881, in a translation of a Japanese dictionary of philosophical terms.②

To sum up, looking up *ziran* in a good dictionary has revealed that it did not mean "nature" until a little over a century ago. If we look up the English word "nature" in the *Oxford English Dictionary*, we learn that it did not have that sense of *daziran* until about 1400 (sense 11). Using the two dictionaries has reminded us that concepts have histories. Some of you will agree with me, perhaps, that a comparative history of the two words would make an interesting research project.

---

① Of the three examples given under sense (1), all are too vague to conclude that they mean 天然,. On the other hand, 非人为地 is appropriate. That is because the word means literally something that happens or exists 然 without something else causing it 自. I understand this usage as not different from sense 3. I have not found *ziran* in a sense that clearly means "natural" until after the Tang period.

② The study by Michael Lackner and his associates has not yet been published; consult Lackner & Vittinghoff 1994. See the discussion in Lloyd & Sivin 2002, 199 – 200.

Body

In the last example, we looked at a Chinese word. Now let us look at an English word, "body." Its main meaning is the entire structure of a human being, animal, or plant. There are several words in ancient Chinese that are often translated "body," but none of them means the entire structure of a human being, animal, or plant, and they differ considerably from each other in meaning.

Let me begin with *shen*, the everyday Chinese word for body. When Zengzi, a disciple of Confucius, said that every day he examined his *shen* on three matters, his point was not looking at himself in a mirror. He actually asked himself "whether in advising other people I have been unfaithful; whether in interaction with friends I have been unreliable; and whether I have practiced what has been taught to me" (Analects, 1.4). His *shen* was a kind of conscience. Today, *shenfen* is the Chinese bureaucratic term for "identity." For 1500 years *shenfen* has meant social origins and status. It has nothing to do with how tall or strong you are. In other words, the meaning of *shen* is much wider than the English word "body," and includes many ideas that have nothing to do with the physical body.

Other old Chinese words that Europeans translate as "body" also have boundaries that differ greatly from those of "body." Three of them are *ti*, *xing*, and *qu*. I have given examples of their meanings in the appendix to this paper. You can see there that *ti* often means an embodiment of some moral quality. *Xing* often means "characteristics" that are not physical. *Qu* often means not "body" but "life."

Even in medical writing, the meanings of these four words do not separate the physical body from various non-physical matters—mental, spiritual, and social—in the way that "body" does. A Chinese two thousand years ago could make it perfectly clear, when he wanted to, that he was talking about a wound in the skin rather than moral ideas. But in his *shen*, unlike the body that the Western physician sees, skin, vitality, emotion, and ethics were equally important parts of one thing. What the Chinese words talk about is not the same as the Western body. That is why I once gave a lecture in Singapore named "Why Chinese Did Not Have Bodies."

Now why were Chinese bodies so different from those of Europeans? Part of the answer comes from the syntheses that formed after about 250 B. C. As I explained in the third lecture, both a well-governed state and a healthy body are in harmony with the regular cycles of the universe. That universe was more than just physical. It was also an organism and a moral order. So was the state and so was the human body.

As part of this relationship, until modern times an important aspect of the *shen* has remained ideals and relationships with other people. It is easy to overlook these characteristics, but comparison makes them stand out. They are the products of a very special historic process that occurred only in China.

East Asian Uses of Astronomy and Medicine

Some European historians have argued that the slow acceptance of Western science by Chinese and Japanese was due to xenophobia. That is not a well-informed judgment. If we look separately at two aspects of science, say astronomy and medicine, it is obvious that Chinese and Japanese responded to them in very different ways. To understand the differences, we have to look both at technical work and at social organization.

Jesuit missionaries from Europe arrived in Japan and China a little before 1600. Their aim was to convert the ruling classes of both countries to Christianity. They quickly found that most such people were not attracted to foreign religions. To attract their attention and convince them that Westerners were a kind of scholar, they wrote books about astronomy, medicine, and other technical topics. The result in 17th-century China was great interest in foreign writings about astronomy; but medical writings had no influence at all. On the other hand, Japanese showed strong interest in Western medical writings, and no interest in astronomy. I will compare the circumstances in the two societies that made their responses so different.

China. In the late Ming period in China, a few high officials such as the famous Xu Guangqi (1562 – 1633) introduced the missionaries' mathematical and astronomical writings into the imperial palace. The emperor refused to appoint the foreign experts to the Directorate of Astronomy, or to publish their writings. Before the beginning of the Qing dynasty, only a very few Chinese who were not officials studied foreign astronomy and wrote about its strengths and weaknesses. Their writings are extremely interesting, but had almost no influence on other people.

When the Manchus invaded China, the missionaries helped them to cast cannon. They convinced the Manchus that the missionaries could be helpful in ruling the Chinese. Within two weeks of the time the Manchu army marched into Beijing, the new rulers showed their gratitude by appointing a European to direct the Chinese Directorate of Astronomy. The new director discarded most of the traditional methods for doing astronomical work, using European methods instead. We can conclude that the Jesuits' astronomy influenced Chinese because the imperial government officially adopted the European methods. ①

There was no such response to the Jesuits' medical writings. At least one book on anatomy, *An Outline of the Human Body from the West*, appeared before the Manchu invasion. But not many copies were printed, and it was not widely distributed. Another reason is that European medicine about 1640 was not superior to Chinese practice. Anatomy was its most novel part, but at the time it had no practical applications in Western medicine, and Chinese physicians had no practical use for it. ②

---

① On Western influence in China and Chinese influence in Europe, see the many writings of Han Qi.
② Ma Boying et al. 1993, 281 – 286.

Japan. The situation in Japan was entirely different. With respect to medicine, we have to understand that country's social structure. Japanese rulers tried to maintain a society in which all work was hereditary, and no one could change occupation. If your father was a cook, you were expected to become a cook and pass down his skills. This was a remarkable structure for a complex society. It was also a difficult structure to live within. You could not legally stop cooking and take up a different occupation.

When this rigid system was put in place, there was no occupation of physician. Therapy was done by various people, many of them priests. Chinese medicine entered Japan early, in the 7th century, but for a long time it was used mainly in the court of the emperor and the headquarters of officials. It slowly spread through society only after about 1400. By 1700 there were enough doctors to treat ordinary people. Not long after, small amounts of Western medical knowledge began to spread from Dutch traders. In 1774, a group of Japanese published an illustrated book on anatomy, translated from Dutch (Sugita 1815).

This book was extremely influential. It had no practical use in therapy, but in Japan it was socially very important. For more than a century, as foreign medical knowledge entered the country, people found that they could create a new occupation. Becoming a doctor might have become illegal, but the new medicine was performing a valuable service, and the government tolerated it. A farmer from the countryside could go to a city, study medicine from a teacher, and call himself a

Using Comparison

N. Sivin
2009. 4. 16

 A good deal of historical work compares something in two different cultures, but one can also compare an idea or custom in the same culture at different times or different places. The meanings of *ziran* and nature, or *shenti* and body, differed. The situations of Western astronomy and medicine in China and in Japan were entirely different. This case suggests that to understand attitudes about foreign science, changes in government, the value of old and new methods, the organization of society, and motivations for learning may be crucial to understanding. Late nineteenth-century Chinese attitudes to the steamboat and steam-driven railways greatly differed. The reasons indicate that Chinese officials of the time recognized the social impact of network technologies and were skilled in diplomacy.

doctor. By the time he retired, he might be rich and have many disciples. Actually, before 1800 he probably would not know much about Western therapy, but having studied anatomy he was as well qualified as anyone other beginner, and his learning gave him authority. He could begin applying what techniques he knew. European medicine became popular in Japan because it opened a career for people who otherwise would have no choice.

But astronomy in Japan was a different matter. The imperial palace used Chinese astronomy, and had a court astronomer. Japanese aristocrats were interested less in an accurate calendar than in astrology and divination, so the hereditary officials in charge (*Onmyōji*) were not highly expert in astronomy. Even in the military government of the 17th century, which employed at least one first-rate mathematician for astronomical work, there was no reason to invent new methods. The Japanese government continued to use the Chinese Season-Granting system (*Shoushili*, 1280) long after China replaced it. There was, in other words, no incentive in Japan to be curious about foreign astronomy.

This example suggests that if we want to understand attitudes about foreign science, it is useless to look at attitudes—or texts—alone. Comparing China and Japan reminds us that such matters as changes in government, the value of old and new methods, the organization of society, and motivations for learning may be crucial. To make valuable comparisons, it is advisable to be aware of the social, political, administrative, intellectual, economic, and other dimensions of history. I will return to this point in my next lecture.

Networks

I was once asked why the Chinese people, unlike Europeans, rejected steam power in the 19th century. I worked with a former student from Wuhan, Zhang Zhong, who earlier did a doctoral dissertation on such questions.① This is not the same issue as in the last example. Our conclusions suggest that it is essential to ask the right technical questions.

Some Western historians have argued that reactionary Chinese institutions, ideologies and habits of thinking made it impossible to accept modern technology before the 20th century. This argument assumes that rational people would have been eager to change if tradition did not stop them. It also assumes that since Chinese were not eager, they must not have been irrational. That did not seem to me a sound explanation. If we consider how complicated international politics in the late 19th century was, it is a good idea to carefully investigate what was rational and what was not.

Political circumstances. It is important to remember that by 1865, because of the Taiping Rebellion and other rebellions, and the Opium Wars, the Manchu government had lost much of its power and much of its income. China did not experience even one year of peace and security from 1840 until 1949. In the second half

---

① Sivin & Zhang 2004, Zhang 1989.

of the 19th century, after China lost the Opium Wars, its law no longer governed parts of its own domains, and the foreign powers could force the government to do anything by threatening another war. On the other hand, they could not make such threats freely. They were not prepared to conquer and govern China; they wanted reliable commercial income. Diplomacy and war were means to be used when trade was threatened, but they had to be used in a way that did not threaten trade even more. So it was possible for intelligent Chinese diplomats to prevent European powers from uniting.

Chinese did not all agree about foreign science and technology—or about anything else. Even high officials disagreed on almost every issue. On the one hand, many members of the imperial family and a number of high Han officials were opposed to accepting anything Western. They realized that Western values, institutions and machines would damage Chinese culture and their own authority.

But others realized that there were important ideas behind European technology. Powerful officials such as Zeng Guofan (1811 - 72), Li Hongzhang (1823 - 1901), Zuo Zongtang (1812 - 85) and Zhang Zhidong (1837 - 1909) learned from the Opium Wars that the only possible way to defend themselves against foreigners was with foreign technology. From about 1860 on, they did buy foreign machines and hired Europeans to build them and to train their users. Li Hongzhang argued in 1874 that the railway and telegraph were essential for defense against naval attack, and the Court gave him permission to build lines for military purposes.

Steam power. You will recall that I had been asked about steam power. We found that Chinese officials responded in entirely different ways to different kinds of steam engines. Before 1875, they eagerly bought and began to build steamboats. On the other hand, they firmly resisted great pressure to let Europeans construct railways. Government policies treated steam locomotives not like ships but like telegraph lines. That seemed strange, but Zhang Zhong found an explanation.

Network Technologies. The historian of technology Thomas Hughes (1983) showed in his book *Networks of Power* that although railways use the power of steam for long-distance transportation, their process of development has nothing in common with those of steamships. On the other hand, it is very much like that of the telegraph, as well as of the electrical power networks that came along later.

Railways are a network technology. They spread their steel tracks in a continuous and highly visible way. They occupy land, so troublemakers can easily stop them. If someone breaks the tracks, trains cannot run on them. They require standard track sizes. As a line becomes longer, its cost goes down. A builder can always make more profit by extending the network.

European businessmen in China understood all of that. It did not take Li Hongzhang and others long to recognize the special characteristics of the network technologies, and to make decisions based on that understanding. Let us look at an example.

The foreign merchants knew that the most important step was building a railway line a few hundred meters long. They were sure that, once it existed, Chinese could not stop them from making it longer. The Chinese officials who dealt with foreigners quickly realized that once a network began, the could not stop it from spreading. From the viewpoint of most Chinese officials, railways and telegraphs owned by foreigners were a potential means of invasion. That was entirely correct.

In 1876, an American diplomatic official organized a company to build what he said would be a road for carriages. When it was finished, it turned out to be a railway about a kilometer long, with a small locomotive. A few months later the owners extended the line to cover six kilometers, and gave it a larger engine, Since the government could neither destroy it nor stop its further spread without danger of another war, it bought it for a very high price and moved it to the island of Taiwan.

Because a railway was not needed in Taiwan then, historians have often described this decision as irrational. But in Taiwan at that time, a network was unlikely to fall under foreign control and spread. In any case, it would be useless for invasion. The merchants realized that they could not outwit their Chinese opponents, and stopped trying to build new networks until after 1895. In that year, a military defeat by Japan took away the last of China's defenses.

In this example, understanding that steam power is not a single kind of technology made it possible to understand the reasoning of Chinese officials in the late 19th century, and to appreciate their ability to do the best they could in the circumstances of the time. To sum up, comparing the uses of steam power in China and the West has led to a better understanding of Chinese skill in diplomacy.

Conclusion

I have given several examples of how comparison, when done thoughtfully and critically, can be useful in solving difficult historical problems. The key is understanding all the circumstances. That makes it possible to learn why certain aspects were similar and others were different. Being able to explain the results of comparison makes it truly useful.

My next lecture will be on cultural manifolds, another tool for making more reliable evaluations.

Appendix

Some Chinese Words for "Body" and Other Things

WORD	MAIN MEANINGS	EXAMPLE OF WIDER USAGE
shen 身	Body, self, personality, identity	吾日三省吾身,为人谋而不忠乎,与朋友交而不信乎,传不习乎。Each day I examine myself on three things: whether in counseling others I have been unfaithful; whether in give and take with friends I have been unreliable; and whether I have practiced what has been taught to me (*Analects*, 1.4). For a medical usage see *xing* below.

(continue)

WORD	MAIN MEANINGS	EXAMPLE OF WIDER USAGE
ti 体	Body, embodiment of some moral quality, personification	阴阳合德而刚柔有体。When the powers of yin and yang merge, hard and soft have their embodiments (*Book of Changes*, *Xici*, B)
xing 形	Form, shape, characteristics, outline of body, body as form	金形之人……其为人……身清廉。The man with characteristics of the Phase Metal... In his conduct... he is pure (*Inner Canon*, *Ling shu*, 64.1.5)
qu 躯	Physical body, person, life	其为人也,小有才,未闻君子之大道也,则足以杀其躯而已矣。He was a man with a little talent who never heard about the great way of the lordly man; that was only enough for him to bring his life to an end (*Mencius*, 7B. 29)

References

Hughes, Thomas Parke. 1983. *Networks of Power: Electrification in Western Society*, 1880–1930. Baltimore: Johns Hopkins University Press.

Lackner, Michael; Natascha Vittinghoff, editors. 2004. *Mapping Meanings: the Field of New Learning in Late Qing China*. Sinica Leidensia, v. 64. Leiden: Brill.

Lloyd, G. E. R., and Nathan Sivin. 2002. *The Way and the Word. Science and Medicine in Early China and Greece*. New Haven: Yale University Press.

Ma Boying; Gao Xi; Hong Zhongli. 1993. *Zhongwai yixue wenhua jiaoliu shi—Zhongwai yixue kuawenhua chuantong* (Cultural contacts between

Chinese and foreign medicine. Transcultural traditions involving Chinese and foreign medicine). Shanghai: Wenhui Chubanshe.

Sivin, Nathan, & Z. John Zhang. 2004. "Railways in China, 1860 – 1898. The Historical Issues." *History of Technology*, 25: 203 – 210.

Sugita Genpaku. 1815/1969. *Dawn of Western Science in Japan. Rangaku Kotohajime*, tr. Ryôzô Matsumoto. Tokyo: The Hokuseido Press.

Sun Xiaochun & Zeng Xiongsheng, editors. 2007. *Songdai guojia wenhua zhong de kexue* (Science and the state in the Song dynasty). Zhongguo Kexue Jishu Chubanshe. Eighteen papers from an International Symposium on the Song State and Science, Hangzhou, 2006.

Zhang Zhong. 1989. "The Transfer of Network Technologies to China: 1860 – 1898." Ph. D. dissertation, History and Sociology of Science, University of Pennsylvania.

5

Using Cultural Manifolds

N. Sivin
2009. 4. 20

　　The approach of cultural manifolds is to look at all the relevant data using all the relevant disciplines to understand questions in the humanities and social sciences. It takes a given problem in all its dimensions, such as science, ideas, social relations, economics, religion, politics, kinship, and others, that are relevant to the problem. Using the idea of cultural manifolds, we find that there was little foreign influence on the *Season-Granting System* because Kublai preferred to keep knowledge from different ethnic groups separate, so that he himself could choose among them. Among the polymaths of the Northern Song period, their breadth of interest was due less to disinterested curiosity than to their official responsibilities. In the early Qing period, most polymaths who were quite innovative scientifically belonged to the evidential research movement. Hundreds of mutually incompatible answers to the "Needham Question" have led to no useful conclusions; its prominence has been due to the narrow technical viewpoint that has encouraged confusion between the absence of a scientific revolution and the absence of a social revolution. Once one begins to examine all the dimensions of these and other problems, it becomes possible to explain the reasons behind them.

I would like to talk today about cultural manifolds(文化整体).① Some of you may find it useful as an approach to historical study. Some scholars in China are already using this method, including Sun Xiaochun in Beijing and Dong Yuyu in Shanghai..② Its meaning is actually simple.

Most scholars are trained as specialists, experts in social history, intellectual history, or some other discipline. By restricting themselves to one dimension of history, they try to reach the greatest possible depth. Such concentration produces most of the scholarship that we read.

In addition to narrow studies written for other specialists, there is also a need for work that covers a wide area of study and communicates with people in more than one specialty. In particular, many questions cannot be answered by looking only at scientific techniques, or only at social relations. Sometimes it is necessary to examine both; sometimes a different discipline, such as economics or religion, is essential. The idea of cultural manifolds is to look at all *the relevant data using all the relevant disciplines*.

Disciplines are not divisions of historical reality; they are divisions of universities. If you are a historian of physics, you

---

① Professor Ren Dingcheng has proposed 文化簇 as an alternative translation.
This lecture is respectfully dedicated to the memory of Professor Bo Shuren.
② See, for instance, Sun & Zeng 2007 for papers by both, and by Dong.

have probably been taught that physical science is what a specialist studies, and everything else is context.① But in historical reality, everything is equally real, and nothing is merely context.

I call the method described in this lecture "cultural manifolds" because in mathematics a manifold occurs in many dimensions. My topic is not mathematics, however, but history. What we are interested in is all the dimensions (ideas, social relations, economics, religion, politics, kinship, and others) that are relevant to a given problem and to the culture in which it occurs. This is not an easy idea to translate into the Chinese name of a method. It might be simpler to describe it as *lishi duofa*. That is clear, but it does not explain why a scholar would want to use it. It also gives the wrong idea that it is useful only for doing history. Actually, it is a wide approach that can be used in any scholarship in the humanities or social sciences. Some people might consider it a kind of cultural history. That may be true, but I don't know. The term "cultural history" is so vague that it is difficult to be certain. The use of cultural manifolds is not new in general historical studies, but not much work on Chinese science and medicine has taken advantage of it. I will give a few examples of its usefulness for that purpose.

---

① *Shangxiawen* is not a very satisfactory translation, but in English the word is used in a wider sense—all relevant outside matters—than in Chinese.

Islamic Influence on Chinese Astronomy

First, there is an interesting problem that came up in my recent book on the Season-Granting System (*Shoushi li*). This was the most advanced system of mathematical astronomy in Chinese history.① The system was completed in 1280 to celebrate the victory of the Mongols over South China after a long and destructive war. From the time that the emperor Khubilai was a young man, he had a group of Han advisors from whom he learned to understand Chinese culture. He usually selected his advisors because of their expertise in what he called *yinyang*, by which he meant astronomy, astrology and divination.

He wanted to persuade the government in South China to surrender. He aimed to rule the reunited China in a peaceful way that would make it easy to collect taxes. His advisors persuaded him that it was essential to design a government of the kind that Han people understood and would accept. Liu Bingzhong (1216 – 1274) and his other advisors built a capital in Beijing and filled it with Chinese-style government offices. They designed a system of state rituals based on those of earlier dynasties. From the viewpoint of Khubilai's Chinese subjects, the government was a traditional one.

Actually, China under the Mongols was divided into four

---

① Sivin 2008. Please refer to this book for detailed notes and references.

legal levels. The Mongols themselves, the rulers, were at the top. Aristocrats were extremely powerful. They could ask the government for Han Chinese to serve as slaves, or for farms with peasant households. Because Mongols generally could not read and write, their administration relied on a second level of non-Han people, what they called *se-mu jen*. They were educated Uigurs and others from Central Asia, Western Asia, and the Middle East, with a few from Europe. They kept records, collected taxes, and did other administration work. In the third class were Han Chinese and others whom the Mongols had conquered early, in North China. The people of the south, who became subjects of the Mongols only in 1276, were the lowest of the low. The Chinese-style bureaucracy had no power over the first two levels. It could govern only Chinese, mainly those in the Metropolitan Province ("the belly," *fuli*), the large area surrounding the capital. The Chinese officials could overcome opposition of Mongols only when the emperor approved their proposals. In other words, although from below the government looked Chinese, that was an illusion. From the top down it was Mongol.

With that in mind, the problem I wanted to investigate was this. China in the Yuan period was part of a very large international empire. Khubilai ruled over only part of it. His government employed Muslim astronomers from Iran and Central Asia, and even had a Muslim Directorate of Astronomy (Huihui sitian jian) in Beijing. The government combined it and the Han Chinese Directorate of Astronomy (Han'er sitian jian) in 1275.

It is natural to expect that, when the work on the new system began in 1276, it would be well informed about Muslim astronomy. Historians therefore have looked for Muslim influences on the work of Guo Shoujing (1231 – 1316) and his colleagues. They found some very small influences, but no important ones at all. The problem is why not?

Most of the historians of Chinese astronomy in China, Japan, and the West have expected to find important influences, and have tried very hard to find them. They have made many proposals, which there is no time to discuss today.① The most rigorous proposals were made by Bo Shuren.② The influences he suggested were extremely small and unimportant.

When we ask why that was true, there are several possible answers. One is that Chinese did not take foreign learning seriously. That idea is easily proved wrong. Another possible answer is that, despite the combination of the Muslim and Chinese Directorates of Astronomy, neither one played an important part in the reform. That, it turns out, is quite true. Khubilai turned over the reform to a new organization, which came to be called the Astrological Commission (T'ai-shih yuan). Its leading astronomers did not come from the old organizations. In fact they were mostly Khubilai's own advisors, all of whom were expert in astronomy, astrology, and divination. The Han Chinese Directorate of Astronomy was

---

① I have analyzed the most interesting of these proposals in Sivin 2008, 218 – 25.
② Bo 1982; see also Chen Meidong 2003, 528 – 29.

turned into a school for low-level workers in the Commission. So far as we know, the Muslim and Han officials in the Directorate played no direct part in the reform. It was common in the astronomical organizations of imperial China—actually, throughout the civil service—for low-level career officials to do little work and to obstruct anyone who was talented and hardworking.① Khubilai created the new Commission to protect his experts from that kind of interference.

But if there were Muslim astronomers in the capital, why did the astronomers who were involved in the reform not learn from them? If the problem was not psychological, what was it? After considering a number of answers to this question, I finally realized that I had been unable to answer it because I thought the problem was astronomical, or scientific. It was not difficult to answer once I investigated the political dimension.

The reason for the lack of influence turns out to be a matter of the Mongol rulers' governing policies. If you read the imperial cookbook of the early Yuan period, *Yinshan zheng yao*, usually described as the most famous Chinese cookbook, you will find many recipes for dishes like boiled wolf's head and similar dishes. You can quickly recognize that that most of the recipes are not for Chinese food. The book contains recipes from many countries that the Mongols conquered (Buell & Anderson 2000). The victors liked to draw on them all rather than turning them into a single style of cooking.

In the same way, Khubilai was a great enthusiast of

---

① For a famous example see Dong 2004.

astrology and divination. He liked to have separate divinations using different systems, so that he could make the choice between them himself. He wanted variety in predictions just as he and other Mongols wanted variety in food.

As many of you know, Khubilai's advisors urged him to revive the imperial examinations, which had stopped at the beginning of the Yuan dynasty. He refused to do that, because he also liked to choose his own officials, or to approve the choices of others, one by one.

That is why Khubilai prevented his diverse subjects from many ethnic groups from exchanging information with each other. Of course, he did not want them to join forces against his rule. But just as important, he wanted to choose himself from the many possibilities that more than one system of astronomy presented. He preferred to keep knowledge separate, by preventing cooperation and exchange between Islamic and Chinese astronomers.

At the very beginning of the Ming dynasty, the new ruler learned that, in the capital, there were a large number of astronomical books in Arabic or Persian. He immediately gave orders for scholars to examine and identify them. He then ordered that they be translated into Chinese, and told Han astronomers to read and learn from them. In fact, we still have translations of two or three of them (Yabuuti 1997, Dalen & Yano 1998, Dalen 2002). What was required for Muslim and Chinese astronomers to communicate with each other was a new dynasty.

Polymaths

I have already mentioned in earlier lectures one of the most interesting figures in the history of science, Shen Kuo (or Gua, 1031 – 1095). One of the things that makes him interesting is the remarkable breadth of his interests (Sivin 1973). The subjects about which he had original ideas ranged from geology to astronomy to mapmaking to linguistics, and on through ritual, music, diplomacy, military fortification, medicine, painting, poetry, tea, and any number of others. I concluded that we can explain this scope only when we recognize that in his life politics, civil service, personal experience, and technical skills were inseparable.

He was not unique. There were other scholar-officials with very wide breadth in the Northern Song dynasty. Anyone who has studied the history of science will recognize the names of Yan Su (fl. 1016) and Su Song (1020 – 1101). In the Northern Song period, these two, along with Shen Kuo, are only the most famous among a number of innovative scholars with very broad curiosities (Zhang Yinlin 1956, Zhuang Tianquan 1993). From that period on, scholars tended to be learned in many fields, from ancient classics to painting. In the Northern Song period, their interests often included science and technology. But in the Southern Song, Yuan, and early Ming periods, what were considered scholars of intellectual breadth rarely included science and technology in their writings.

In North China under the Mongols, as we have already

seen, there was another important change. Many young scholars studied astronomy, astrology, mathematics, and divination. There were no imperial examinations in the early Yuan period, so such studies offered one of the best opportunities to earn a living. When the Mongols revived the regular civil service examinations in 1315, that changed the situation again. But although many Chinese earned a living as diviners, we do not find many new books on science.

In certain periods of the Ming and early Qing dynasty, questions on observational and computational astronomy and other technical topics appeared regularly in the imperial examinations. This forced every elite young man to study these fields. Remarkably, it also did not cause a rebirth of scientific activity or writing among graduates (Elman 2000, 2005).

There was another important change from the end of the Ming to the late Qing period. A limited number of scholars in what became the "evidentiary research" (*kaozheng*) movement took up the study of mathematics, astronomy, medicine and other fields. Beginning with the leadership of Gu Yanwu (1613 – 1682), such study flourished until the mid-Qing period. Gu Yanwu, the outstanding astronomer Mei Wending (1633 – 1721) in the early Qing period, and later on Dai Zhen (1724 – 1777), and others argued that to understand the original meaning of the ancient classics it was necessary to use the tools of philology and the sciences (Elman 1984). This group was quite innovative in many ways, especially in putting together Chinese and Western mathematical techniques.

To sum up, if we investigate the question of when and

why scholars with broad interests included scientific work among their writings, we find that this happened frequently in the Northern Song period, and from the end of the Ming period on. Although large numbers of young men studied science at other times because they hoped it would help them earn a respectable living, that did not lead them into real scientific activity. Why not? Keep in mind that the study of science before modern times usually meant memorizing scientific classics. That was a good way of collecting knowledge, but it did not naturally lead large numbers of scholars into programs of scientific research.

On the other hand, in the Northern Song period, what scientifically-minded scholars had in common was official responsibilities connected with the sciences. Shen Kuo was mainly a financial official, but at one time he was the supervisor of the imperial astronomical bureau. Yan Su made instruments for the imperial palace, and Su Song (who became prime minister) was involved in an astronomical reform. Because the Northern Song government became much more deeply involved in problems of public health than any earlier government, it was natural for officials to study medicine.

The evidential research scholars were generally from Jiangnan. Many of them taught in schools that in many ways were like modern universities. They had a strong sense of mission. Gu Yanwu and others argued that the reason the Ming dynasty fell to foreign invasion was that it had lost the true teaching of Confucius and other sages. The reason, they believed, was that scholars had contaminated those teachings with heterodox doctrines that came from Buddhism and Daoism. It was therefore essential to recover the true original teachings, ignoring the contaminated scholarship of many centuries and returning to the original texts. Their tools, philology and scientific scholarship, would make that possible. So in the case of the evidential research movement, it was the dedication to this aim by a number of scholars who were mostly not officials but famous teachers that led to an important role in the history of science.

To sum up, the role of science as an activity of broadly educated scholars changed several times from the Northern Song period on. If we examine the cultural manifold of this problem, we can see that different dimensions played important roles in explaining the reasons for these changes. First, the technical problems were always important; this is not a matter of internal versus external history. In addition, the political factors changed considerably from time to time; patterns of livelihood also changed; if we knew more about the economics of research we would no doubt find it played an important role. To fully understand this problem will require considering many dimensions of intellectual, personal and social identity.

Now let me give a couple of examples of comparative issues that are ripe for investigation.

The "Needham Question"

Historians of science have all thought critically about what has come to be called the "Scientific Revolution Question" or the "Needham Problem." Most people who have discussed this problem overlooked the most important first step: defining the term "scientific revolution." Everyone knows that the modern historical understanding of this term comes from Thomas Kuhn's *The Structure of Scientific Revolutions* (Kuhn 1962; see also Hollinger 1973). A scientific revolution is a special kind of change that involves new methods, a new understanding of what a scientific problem is, a new understanding of how to solve a scientific problem, and even a new understanding of what a solution to a scientific problem is. When these changes take place, it may be due to an entirely new theory, or a theory that comes from outside the culture (Kuhn 1962; see also Hollinger 1973). The first phase of the Copernican Revolution, for instance, took place more than 70 years after Copernicus' death, in Italy, due to the writings of Galileo. Copernicus, in Poland, had used his cosmological ideas only to improve Ptolemy's ancient astronomy, but Galileo used them to attack Aristotle's universal philosophy, which was the foundation of European university education. The scientific revolution was not reinvented in France, but was imported into France from Italy. Its importation into China was not different from its importation into various European countries. In fact, the scientific revolution began to enter China at about the same time it entered France, and earlier than it entered many European countries.

Some historians have denied that there was any revolutionary change in Chinese science. But if we look at the work of the best Chinese astronomers, such as Wang Xishan (or Xichan, 1628 - 1682) and Xue Fengzuo (ca. 1620 - 1680), we find that they responded not only quickly but critically to the information about mathematical and astronomical techniques that they read in the writings of the Jesuit missionaries. Many historians did not realize this, because they assumed scientific revolutions must have the same social consequences that they had in Europe. In China it is clear that the new scientific ideas did not have wide social consequences. Many historians therefore believed that Chinese scientists were not able to respond to revolutionary scientific ideas. But that is because they did not read the writings of Wang Xishan, Xue Fengzuo, and the other innovative astronomers. When you read their work, the obvious conclusion is that revolutionary scientific change can take place without causing social change. And if that is true of China, it is true everywhere.

In other words, what is abnormal is the social revolution that eventually swept over Europe after its Scientific Revolution. In fact, many specialists have shown that while some Europeans used new scientific ideas to encourage social transformation, others used them to block it (Jacob 1976, 1981; Shapin 1996).

The historical question does not lie in China: why did the social revolution happen in Europe? But that is not a question that historians of Chinese science can answer.

I encouraged my colleagues to examine systematically what did happen after Chinese began to think about Western science.

Some, especially scholars in China, have done so (Han Qi 1999, Huang Yinong 1990). Other scholars are still confused because there was no social revolution in China, so they think that a scientific revolution was impossible. They usually assume that the Chinese scientists were not clever enough, or were too rigid, to make productive use of Western science until the twentieth century. They therefore continue to guess answers to the "Needham Question." The result is hundreds of different and mutually incompatible answers. They have led to no useful conclusions (Liu & Wang 2002). I hope that scholars will not waste any more time with useless assumptions. Their effort is badly needed to compare the responses of Chinese astronomers and mathematicians in the 17th century with those of scientists in various European countries.

The prominence of the "Needham Question" has been due to a narrow technical viewpoint. That is why that confusion between the absence of a scientific revolution and the absence of a social revolution has lasted so long. As soon as we realize how important it is to look at both scientific questions and social questions, the confusion is over. In the history of science the polemics over internal and external history ended thirty years ago; there is no longer any need to waste time by doing only one or the other.

Bu' erchen

As I said in the last lecture, in addition to comparing two cultures in the same period of time, it is useful to compare one phenomenon in China at different times. Keep in mind that the most important part of a comparison, once you find the differences, is to explain why they are different. When comparing two different times, this amounts to explaining historic change. In this example I will discuss a historical question that no one has yet answered.

A Chinese custom that affected science was the ancient usage that forbade an official to serve two successive dynasties (*bu' erchen*). Many scholars went even further, refusing to serve a new dynasty even though they were not officials, or even had no prospects of an official career under the old one. This prohibition does not seem to have affected Han Chinese in the Yuan period. The Mongols had no difficulty appointing Chinese astronomical talent to work on the astronomical reform that began in 1276. All of the leaders of this reform were personal advisors of Khubilai. His closest advisor, Liu Bingzhong, came from a family that had served the Jin rulers, but he eagerly went to work for Khubilai, and recruited many other experts in what people at that time called "yinyang"—that is, astronomy, astrology, and divination. After the Southern Song government surrendered in 1276, former Song astronomical officials such as Chen Ding provided the Mongols with scientific experience that no northerner had. He had already taken part in an astronomical reform. He did not choose to die rather than serve, as others had done in previous dynasties.

Now let us look at the transition from the Ming to the Qing, which again was from a Han government to an alien one, namely government by the Manchu people. Many exceptionally able scholars refused civil service appointments, even though they had never been Ming officials. They became teachers of astronomy and mathematics, and physicians, instead. Mei Wending, Wang Xishan, and Xue Fengzuo, the best astronomers of their time, are examples. Wang Xishan, who was only 16 years old at the beginning of the Qing dynasty, tried to drown himself. Someone rescued him, but for the rest of the life he supported himself as a teacher. He wrote dates using the reign years of the Ming dynasty. Since that was a serious crime, he always wrote in seal script, so few people could read his handwriting.

Fu Shan (1607 - 1684), was an outstanding medical practitioner and author, and one of the very best calligraphers and painters of his time. In the Ming period, he never became an official. The Qing court offered him official appointments, and pressed him very hard to accept them. When he said he was too ill to go to the capital, the government ordered that his bed be carried to the palace, so he could accept an official post without getting out of bed. When he arrived in Beijing, he announced that he would kill himself if he was forced to accept even an honorary appointment.① Many Qing officials asked him to write calligraphy or paint for them, but he refused. For the rest of his life he was extremely poor. He supported himself and his son by selling medicines, not to officials but to ordinary people.

We have seen a great difference between the Song-Yuan and Ming-Qing transitions—the idea of *bu' erchen* was unimportant in the first, and prevalent in the second. What accounts for this contrast, which affected science? I have not thoroughly studied this question, so I have no idea. Looking at all the dimensions of this concept through history can tell us. All it takes is hard work, thoughtful analysis, imagination, and an open mind. Perhaps someone at this lecture can solve this fascinating problem. But what is most important is the questions that you are already curious about.

Conclusion

I have given a number of examples of cultural manifolds in this lecture. The problem of Islamic influence on Han astronomy turned out to be a matter, not of scientific factors, but of political ones. The other examples were comparative.

---

① See the biographies of Wang and Mei in Jin Qiupeng 1998, pp. 660 – 675. For a biography of Wang in English, see Sivin 1995. On Fu, see He Shuhou 1981. There is no substantial biography of Xue, but see his writings in *Lixue huitong* (Eclectic astronomy, preface 1662); some versions carry the synonymous title *Tianxue huitong*. For the complicated bibliography of these two titles, see Sivin 1995, pp. 28 – 29, note 39.

The problem of the Needham question compared the Scientific Revolution in different countries. The problems of the broad interests of scholars, and of scholars who refused to serve two dynasties, both compared different periods in Chinese history.

All of these examples illustrate a simple point. Although it is a custom in universities to train specialists in one kind of historical analysis or another—in intellectual history or social history, for example—this narrowness makes it impossible to find reliable solutions to many historical problems. The idea of cultural manifolds is useful simply as a reminder of this.

But if you are being trained as a specialist, how can you become a generalist?

As you begin doing research, your teachers will advise you to work on small topics so that you can master techniques. Eventually, writing a dissertation demonstrates in a large project that you have learned the techniques of your special field. From then on, you can decide to study new techniques and methods of analysis of other specialties. You can choose to design projects that will look at more dimensions of a problem. By extending your experience in different specialties, you can eventually use cultural manifolds to understand topics that cannot be understood by using a narrower approach. Your intelligence, motivation, and imagination are badly needed to contribute to the worldwide history of science, technology, and medicine.

References

Ansari, S. M. Razaullah, editor. 2002. *History of Oriental Astronomy. Proceedings of the Joint Discussion 17 at the 23rd General Assembly of the International Astronomical Union, Organized by the Commission 41 (History of Astronomy), Held in Kyoto, August 25 – 26, 1997.* Astrophysics and Space Science Library, 274. Dordrecht: Kluwer Academic Publishers.

Bo Shuren. 1982. "Shi tan you guan Guo Shoujing yiqi de jige xuanan (An attempt to resolve some open questions about Guo Shoujing's astronomical instruments)." *Ziran kexueshi yanjiu*, 1. 4: 320 – 26.

Buell, Paul D.; Eugene N. Anderson. 2000. *A Soup for the Qan: Chinese Dietary Medicine of the Mongol Era as Seen in Hu Szu-hui's Yin-shan Cheng-yao*. The Sir Henry Wellcome Asian Series. London: Kegan Paul International. Historical context on Mongol period, analysis of text, unpunctuated text, translation, notes and reconstructions of recipes, list of ingredients in English, and essay on grain foods of early Turks.

Chen Meidong. 2003. *Zhongguo kexue jishu shi. Tianwenxue juan* (History of Chinese science and technology. Astronomy). Beijing: Kexue Chubanshe.

Dalen, Benno van. 2002. "Islamic Astronomical Tables in China: The Sources for the *Huihui li*." In Ansari 2002, 19 – 31.

Dalen, Bruno van; Michio Yano. 1998. "Islamic Astronomy in China: Two New Sources for the *Huihui li* (Islamic calendar)." In *Highlights of Astronomy*, ed. J. Andersen, CXVIII, 697 – 700.

Dong Yuyu. 2004. "Beisong tianwen guanli yanjiu (A study of astronomical administration in the Northern Sung dynasty)." Ph. D. dissertation, history of Physics, Shanghai Jiaotong Daxue.

Elman, Benjamin A. 1984. *From Philosophy to Philology. Intellectual and*

Social Aspects of Change in Late Imperial China. Harvard East Asian Monographs, 110. Cambridge, MA: Council on East Asian Studies, Harvard University.

Elman, Benjamin A. 2005. *On Their Own Terms: Science in China*, 1550 – 1900. Cambridge: Harvard University Press.

Elman, Benjamin A. 2006. *A Cultural History of Modern Science in China*. New Histories of Science, Technology, and Medicine. Cambridge: Harvard University Press.

Han Qi (1999). "Zhongguo kexue jishu de xi chuan ji qi yingxiang (The westward transmission of Chinese sci-ence and technology and its influence)," in *Dong xue xi jian congshu* (Collection of works dealing with the spread of Eastern Learning to the West). Shijiazhuang: Hebei Renmin Chubanshe.

He Shuhou (1981). *Fu Shan zhuan* (Biography of Fu Shan). Chang'an: Shaanxi Renmin Chubanshe.

Hollinger, David A. 1973. "T. S. Kuhn's Theory of Science and Its Implications for History." *American Historical Review*, 78. 2: 370 – 393.

Huang Yinong. 1990. "Tang Ruowang yu Qing chu xi li zhi zheng-tonghua" (Johann Schall von Bell and the legitimation of Western astronomy at the beginning of the Qing period), in Huang, *Xin bian Zhongguo keji shi. Yanjiang wengao xuan ji* (New history of Chinese science and technology: Selected lecture notes and drafts). 2 vols., Taipei: Yinhe wenhua shiye gongsi, vol. 2, pp. 465 – 491.

Jacob, Margaret C. 1976 *The Newtonians and the English Revolution* 1689 – 1720 Ithaca: Cornell University Press.

Jacob, Margaret C. 1981 *The Radical Enlightenment. Pantheists, Freemasons and Republicans* London: George Allen & Unwin.

Jin Qiupeng (ed.) (1998). *Zhongguo kexue jishu shi. Renwu juan* (History of Chinese science and technology. Persons). Beijing: Kexue chubanshe.

Kuhn, Thomas S. 1962/1970. *The Structure of Scientific Revolutions*. University of Chicago Press. 2d ed., 1970.

Liu Dun & Wang Yangzong (eds.) (2002). *Zhongguo kexue yu kexue geming. Li Yuese nanti ji qi xiangguan wenti yanjiu lunzhu xuan* (Science and scientific revolution in China. Selected discussions of the Needham Problem and related questions). Liaoyang: Liaoning Jiaoyu Chubanshe.

Shapin, Steven (1996). *The Scientific Revolution*. Chicago: University of Chicago Press.

Sivin, Nathan. 1973/1995. "Shen Kua (1031 – 1095)." In *Science in Ancient China. Researches and Reflections*, chapter 3. Aldershot, Hants: Variorum.

Sivin, Nathan. 1995. "Wang Hsi-shan (1628 – 1682)," in *Science in Ancient China*, chapter 5.

Sivin, Nathan. 2008. *Granting the Seasons: The Chinese Astronomical Reform of 1280, With a Study of its Many Dimensions and a Translation of its Records*. Sources and Studies in the History of Mathematics and Physical Sciences. Secaucus, NJ: Springer.

Sun Xiaochun & Zeng Xiongsheng, editors. 2007. *Songdai guojia wenhua zhong de kexue* (Science and the state in the Song dynasty). Zhongguo Kexue Jishu Chubanshe. Eighteen papers from an International Symposium on the Song State and Science, Hangzhou, 2006.

Yabuuti, Kiyosi. 1997. "Islamic Astronomy in China during the Yuan and Ming dynasties," rev. & tr. Benno van Dalen. *Historia scientiarum*, 7. 1: 11 – 43.

Zhang Yinlin (1956). "Yan Su zhuzuo shiji kao" (On the writings and achievements of Yen Su), in Zhang, *Zhang Yinlin wenji* (Collected works of Zhang Yinlin). Taibei: Zhonghua congshu, pp. 116 – 124.

Zhuang Tianquan et al. (eds.) (1993). *Su Song yanjiu wenji. Jinian Su Song shouchuang shui yun yixiangtai jiubai zhou nian* (Collected studies of Su Song to commemorate the 900th anniversary of the water-driven astronomical observatory-clock invented by him). Amoy: Lujiang chubanshe.